James A. Ekin Criswell, Warren Beecher Hutchinson

Patents and How to Make Money Out of Them

James A. Ekin Criswell, Warren Beecher Hutchinson

Patents and How to Make Money Out of Them

ISBN/EAN: 9783744715003

Printed in Europe, USA, Canada, Australia, Japan

Cover: Foto ©berggeist007 / pixelio.de

More available books at **www.hansebooks.com**

PATENTS

AND

HOW TO MAKE MONEY OUT

OF THEM

BY

W. B. HUTCHINSON

OF THE NEW YORK BAR

NEW YORK:

D. VAN NOSTRAND COMPANY

23 MURRAY AND 27 WARREN STREETS

1899

PREFACE

The main object of this book is to tell how to make money out of inventions and patents.

It treats chiefly of the business side of inventions. Its authors have had a large experience in matters relating to patents, and believe that a little honest and reliable advice as to how to invent, to patent, to introduce, to sell and to protect an invention will be appreciated by all who have or are likely to have business in this line. Nearly all the literature on this subject has been in the nature of text-books on the law of patents, which are practically of no use to the business man, or in the form of adroit advertising matter, the object of which has been to transfer dollars from the pocket of the inventor to that of the advertiser. It is a recognized fact that many ingenious men waste their ingenuity by exercising it in the wrong direction. To such men it is hoped this book will be an aid. Others, through ignorance of the nature of patents and the proper method of procedure, fall into the hands of incompetent and unscrupulous attorneys and fail to secure that to which the law entitles them, and so see a competency slip from their hands. To such this book, if followed, will prove a blessing. Again, manufacturers and other business men often meet with loss in purchasing patents which do not cover the inventions to which they relate, or to which the title is in some way defective. To these the book will be of great assistance. Competent lawyers are often unable to advise their clients as to practical means of selling, licensing or introducing a particular invention. It is hoped that this book will be a help to them. Finally, we commend this volume to all people having any connection with patents, and trust that the practical experience which is herein embodied may, in some way, be a help to all.

This book is not intended to take the place of an attorney; it is not published to boom a patent agency; it is not a collection of legal lore and decisions for the especial use of lawyers, but it is intended as a practical guide for inventors, manufacturers, lawyers and business men generally who have anything to do with patents.

We have avoided text-book form in this book and have refrained from using foot notes, as these, to the average reader, are confusing and disconcerting. The reader is asked to take our word for the facts herein, and as to matter of opinion and advice to take it for what it is worth.

NEW YORK, February, 1899.

CONTENTS

BOOK I.

PATENTS GENERALLY.

BOOK II.

PATENT OFFICE PRACTICE, TRADE-MARKS, COPYRIGHTS.

BOOK III.

PATENTS COMMERCIALLY CONSIDERED.

APPENDIX.

PATENTS

AND

HOW TO MAKE MONEY OUT

OF THEM

BOOK I.

CHAPTER I.

ORIGIN AND NATURE OF PATENTS—THE
MODERN APPLICATION OF THE TERM.

Originally patents were monopolies and were more often granted to give the patentee a monopoly in large tracts of land for commercial, mining, manufacturing and general business purposes, such, for example, as the patents granted to the first colonizers in America for immense districts of land, to which the name of the original patentee has, in many cases, been applied.

Often at the present time the Government grants to homesteaders and other persons patents for the land which they have preempted, purchased or otherwise secured. But in its ordinary application a patent is a Government grant giving an inventor the exclusive right to make, sell and use his invention for a term of years. In the United States a patent is granted for seventeen years.

A patent, in its modern acceptation, is not a monopoly, but is a consideration offered as an inducement for a person to invent. In other words, it is a prize or reward for his ingenuity, and gives him the exclusive right to make, use and sell the invention for a limited term in consideration of the benefits of the invention to society.

To secure a valid patent, therefore, the Government requires by law that the inventor shall file in the Patent Office drawings and descriptions of his inventions, sufficiently clear to enable persons skilled in the art to make and practice the invention, so that at the end of the term of the patent the public can have access to the invention through the records and so reap the benefits thereof.

As above remarked, patents were originally monopolies, but in the 21 James I., 1624, the Statute of Monopolies, so called, was passed, by which the granting of special and exclusive privileges in trade were prohibited, but the statute specifically excepted "letters patent and grants of privileges for the term of one and twenty years or under, heretofore made for the sole work or making of any manner of new manufacture within this realm to the first and true inventor or inventors of such manufactures."

This statute marked the beginning of modern

patent laws for the protection of new and useful inventions.

In France the first patent law was passed in 1791. In the United States the patent system has grown up under a positive grant in the Federal Constitution and by reason of Statutes, the first of which was passed in 1790, and others from time to time to the present day.

A patent is wholly a creature of statute, and other nations have been somewhat slow in following the lead of Great Britain, the United States and France, and even at the present time there are some countries which have no patent law. Nearly all the civilized nations, however, recognize the importance of a patent system to foster and encourage inventions, and, at the present day, there are patent laws in nearly every civilized country.

As a rule, the laws are such that an alien may have practically the same protection for his invention as a native, though in many countries it is stipulated that the invention must be worked within a certain specified time or the patent forfeited.

It is contended by many that a provision of this nature should be made in the United States, so as to prevent large corporations from buying and controlling numerous patents for inventions which are never marketed, thus depriving the public of the benefit which it should receive, but it is a question whether the time is ripe for such a change.

It is as true now as of old that "nothing succeeds like success," and it is a fact beyond dispute that the patent system of the United States has brought inventions to a wonderful state of efficiency, so that the country leads the world in valuable improvements.

CHAPTER II.

Visit an oculist and, whatever your bodily ailment, he will probably tell you that the difficulty originates with the eyes. A skilled surgeon will likewise conclude that the supreme remedy for every ill is the knife. So, politically, one set of people will attribute the prosperity of the United States to its natural resources, another to the policy of protection, another to certain financial systems, and so on *ad infinitum.*

But we believe it is a demonstrable fact that the patent system of America has done more to promote its commercial supremacy, its wonderful prosperity and general well-being than any other cause.

The workings of the patent system are quiet and unobtrusive. The inventor does his work, and is, as a rule, comparatively unknown. He is not greeted with the applause of the fighting man or the orator, but he does and has done more for the world than any other man or set of men.

The patent system of America is more liberal to the inventor than that of any other country, and it has been the policy of the Government to do what it could to encourage inventions, notwithstanding the fact that the agricultural communities have usually opposed patents as being in the nature of monopolies.

The men who have made American manufactures famous and have, by their improvements, brought the United States to the front rank as a manufacturing and commercial nation, would not have brought out their improvements and could not have found money to exploit them were it not for the fact that the Government has provided reasonable protection.

People do not work for the love of working. It is not human nature. There must be some sort of encouragement and stimulus. The patent system of the United States has provided this stimulus, and has opened almost the only avenue of success on which the poor man can successfully enter.

It is generally recognized that patent law forms an important branch of American jurisprudence, still the real value of American patents and of the American patent system is appreciated by comparatively few people, and few know the boundless benefits the world has derived from the achievements of American inventors. Many of them have cut niches in the temple of industrial fame that will last forever.

When it is remembered that the richest nation in the world is now the United States, that her improvements and manufactures are fast taking the lead, and that the whole volume of manufacturing business in America is or has been based on patents—that is, that the articles made or the machinery for making them are or have been subject to patent—then the enormous value of patents to the public begins to be appreciated.

The principle of our patent system was early recognized; for instance, before the patent law of 1790 Massachusetts granted in 1786 a cash subsidy to Alexander and Robert Barr, of Scotland, and Mr. Orr, of East Bridgewater, Mass., to encourage the introduction of cotton-manufacturing machinery. This was done to carry into effect the wonderful inventions of Hargreaves and Arkwright. Aid was also granted later to Almy, Brown and Slater, who first manufactured cotton goods in Rhode Island, and subsequently the inventor Lowell, whose monument is one of the most thriving manufacturing cities in the Union, made his first loom, the model being completed in 1812. This he patented, and it marks an epoch in the manufacturing history of America.

Eli Whitney, whose name has for generations been a household word, was a poor lad ignorant of the cotton industry, and yet he invented the cotton gin in 1793, which made it possible to prepare cotton cheaply for manufacturing purposes, to supply the previously invented spinning-jenny and the subsequently improved loom, so that thus early, through the aid of the States and the protection of the patent system, was inaugurated an industry the importance of which can scarcely be realized. The history of the cotton industry is substantially a duplicate of all the other important manufactures of America.

The names of McCormack, Heines, Ketchum, Manny, Wood and others are familiar wherever mowing and reaping machines are known and the inventions relating to mowers and reapers and other agricultural instruments which have been fostered by the patent system would not probably be brought to public notice except for it and by reason of this protection. America now supplies agricultural machinery to the world, and not only that, but is able to produce crops at a price which defies competition and which enables American products to be sent to the four corners of the earth.

The career of Fulton in connection with steam navigation is well known, and the importance of his inventions is understood by every schoolboy. All know how Singer, Howe, Wheeler, Wilson and others have made the sewing-machine known in every hamlet, not only in America, but in Europe; how modern processes, every one of which is or has been patented, have enabled America to ship iron and steel even to Great Britain, as well as to all other parts of the earth; how American locomotives are now rolling through the wildernesses of Siberia and over the mountains of Japan; how American boots and shoes are sold everywhere, and how every one of these industries has been made what it is by reason of improved machinery. Comparatively young people can remember how boots and shoes were made by hand in scattered country districts until the introduction of the McKay machines subsequent to 1860, and immediately thereafter how important cities sprung up because the machinery made it possible to turn out the manufactures cheaply and in small localities.

It is as familiar as A, B, C how Morse and Bell and Edison and Thomson and other lesser lights have made the electric phenomena serviceable; have "harnessed the lightning" and have made what was formerly a superstitious wonder a common vehicle of every-day use.

Most, if not all, of the men referred to above, together with others who are well known, would have lived and died poor were it not for the fact that the patent system has opened to them a laudable source of wealth. It must be remembered that while the patent system has been a blessing to these men, still the chief blessing has, after all, been to the American public, to whom the inventors have turned over the wonders of the nineteenth century, and by whom what were formerly luxuries are now brought to every household.

It cannot be contended that these industries would have been promoted by the expenditure of years of toil on the part of those individuals and the expenditures of vast treasures unless the patent system had offered reward for such endeavors and for such expenditures. From these sources have resulted the wonderful manufacturing conditions in America, and it seems clear that the patent system has been of more value to Americans than any other one thing.

Not only this, but the beneficent effects are felt in war as well as in peace. Every manufacturer and almost every firm now knows the difference between the work of a clodhopper from darkest Russia and that of an intelligent native American. The vast difference is due, in a great measure, to the fact that almost every American is at the present time more or less familiar with mechanics, owing to the wide distribution of mechanical inventions. As long ago as the civil war it was found that every company contained men who could, when occasion required, rig up a locomotive, repair a telegraph line, send telegrams if necessary, run a printing office, or do anything which the occasion demanded. So, likewise, in the recent Spanish war, the great superiority of the American navy is said to lie principally in the fact that it was manned chiefly by mechanics who were familiar with mechanism and who could handle with effect machinery constituting a modern fighting navy.

It is, of course, understood that the part any one of the various industries plays or has played in American development is in many instances subject-matter for a volume itself, and the matter is only referred to here in a general way to illustrate the immensity of American manufactures and the fact that these manufactures have been introduced primarily by the American patent system.

CHAPTER III.

We are apt to forget, surrounded as we are by the many comforts and inventions of the last few generations, that the dawn of the Nineteenth Century found mankind in about the same condition, so far as industrial development is concerned, as he was when the pyramids were built or when Phidias adorned Athens with the artistic treasures which were a copy for posterity.

The difference was merely one of degree. The carrying trade of the world was done in ships which were the same in principle as those the hardy Norsemen navigated in their early trips to Newfoundland and the American Continent, and with which Columbus made his memorable voyage across the Atlantic. The student did not "burn the midnight oil," but with difficulty perused his after-dark studies by the light of a pine knot or a tallow dip. One modern steamship like the "Kaiser Wilhelm Der Grosse" will carry more merchandise in a year than perhaps the whole commerce of America amounted to at the beginning of the century. Steam did not affect the ocean carrying trade, the land carrying trade or passenger traffic; New York and Boston were far distant municipalities; Philadelphia was as far from New York as Denver is now. Communication was so slow and uncertain that only the most important events were attempted to be transmitted and the result was often disastrous. As late as the war of 1812, the most important battle was fought long after the treaty of peace had been agreed to. Our late war with Spain was fought and finished in two hemispheres in less time than it took to get a message to Europe and return. Practically everything consumed was of hand manufacture and mostly homemade.

The great industries of modern times were as yet undreamed of. It is said that the steel output of the United States for 1898 was greater than the steel manufacture from the time of Tubal Cain to the beginning of the Nineteenth Century.

Not one of the great enterprises of the present era had been inaugurated. The standard of living was low. Armies could be moved no quicker than in the days when Hannibal marched his legions from Spain to Italy, or when Julius Cæsar made his wonderful march across the Alps. The telegraph, the telephone, the electric light, the typewriter, the sewing machine, the mowing machine, the locomotive, the modern weaving machinery, in fact nearly everything of common and necessary use in the industrial arts, was as yet undiscovered.

Alfred Russell Wallace, who coöperated with Darwin in formulating the doctrine of the "survival of the fittest" and who is conspicuous as a writer of natural history and a great and scientific observer, says that the Nineteenth Century marks the most important epoch within the whole historic period or, perhaps, since the stone age. He calls it "The Age of Invention" and compares the beginning of this era with the introduction of fire.

To enumerate in these pages the many wonderful inventions or more than hint at their importance would require more space than could be given in a book of this character. The importance of the Nineteenth Century inventions, scientifically, industrially and socially, are well understood. But what is not understood and what has been generally overlooked is the fact that this supremely important period is due largely to the beneficent patent system of the most progressive nations. Naturally we find the most liberal patent system in the country which leads the procession in inventions.

It may be said that the patent system follows inventions, and while it is true that one is dependent on the other, still it is certain that nearly all the inventions which have done so much for the world and which have raised the standard of living and general intelligence would never have been commercially and practically developed were it not for the initial protection of the patent system.

It is not necessary to quote authorities to show that man is inherently selfish, and while he loves approbation, still he would never go to the extent to which most great inventors have had to go, would not have denied himself and his family, would not have labored for years at great expense and at great suffering in some cases, merely to secure the honor of bringing forth a great invention. It has required something in the nature of a pecuniary reward or, at least, something to hold out the hope of reward, to induce the inventor to properly develop his inventions and to exercise his ingenuity to the utmost.

It is common to laud the orator or the great general or some philanthropist far more than the inventor, but the real inventor is the king among men. He does not always invent a machine, but his breadth of mind and his sweep of view comprehend everything between heaven and earth. He fears nothing, not even ridicule, but has a mind open to discover truth wherever it may be found. He does not always invent a machine. He may, like Homer or Milton, exercise his mental characteristics to pour forth songs to delight the ages. He may, like Galileo or Columbus or Copernicus, change the human idea of the universe, or he may, like Fulton or Morse, use his ingenuity to industrially help the race, but in every case the mental attitude and characteristics are the same.

It is to such men that the world should do homage.

Perhaps it may not appear at first view what this has to do with the commercial value of patents, but on an instant's reflection it will be seen that these inventions, which really constitute the modern industrial fabric, have all been the means of colossal fortunes for those interested in them, especially in view of the fact that every great invention, instead of closing the avenue of inventive work along that line, has always opened a field which has been filled immediately by lesser inventions worthy, however, of commercial exploitation and, as a rule, profitable.

The smaller inventions are frequently in fact the most profitable to one whose means are somewhat limited, because they can be developed and exploited for a comparatively small sum, while the larger affairs usually require modest fortunes to show their worth or the lack of it.

The "little things" are often "the big things" in the aggregate.

Everyone knows how fortunes have been quickly made out of glove fasteners, shoe eyelets, and a hundred other "little things." A match is a little thing, but yet the match industry is large enough to absorb the attention of one of the great trusts of the country. And the industry is paying dividends on $18,000,000.

Everyone may not know that even the wooden toothpicks which are apparently so insignificant are made and sold in carload lots, and that all the machines for making them have been patented, while originally the toothpicks themselves were subject to patent. Clothespins, shoe nails and even peg wood for boots are all sold in immense quantities.

Several fortunes were made in the manufacture of paper collars. A good toy will usually realize a fortune for its promoters in a season or two. And so we might go on indefinitely. The point is that, if there is a reasonably large sale for an invention and it is properly protected, it is worthy of attention. If it is not or cannot be patented, it is not ordinarily worthy of attention for competition then reduces the price to practically the cost of the labor and material of which it is made.

A device, machine or process which is patented and which to any appreciable extent decreases the cost of making any staple goods is of self-evident value. Competition is so close that a small saving in cost of goods, or a means of making better goods at a given cost, is of great value, the value depending, of course, on the line of goods and the relative quantity consumed.

Nothing in the manufacturing line is so good as a good patented invention. Suppose, for instance, that the leading manufacturers in a certain line have pooled or formed a trust. Such a trust can easily, by their well-known methods, crush any outsider having only its facilities. But if the outsider gets control of a better article or a cheaper or better machine or process of manufacture, then the trust must make terms with him.

It will be seen that the patent protection offers almost the only means of securing large profits on a reasonable investment so far as ordinary industrial business is concerned, that is, business outside of the great monopolies which have absorbed certain lines of commerce.

There are only a few ways of avoiding this destructive competition. One is by combining or pooling all the industries of a certain kind in the form of a trust and another is to manufacture some articles on which there can be patent protection, or which have become known and favored by the trade and are recognized by a lawful trade-mark.

This fact should make the manufacturers—and does make them—eager to take up a good invention and it should also cause them to be very careful to see that the invention is properly patented so that they can safely enter into its manufacture.

The alert inventor will also strive to invent along practical lines as pointed out in another chapter.

The manufacturer or inventor will likewise be on the lookout for the opportunity which may show itself but once and then momentarily ; and both knowing the many elusive qualities of patent rights and that a patent is a creature of statute, shaped in every instance more or less according to skill, should be careful to see that their interests are properly safeguarded.   This can only be done by those skilled in such matters, and many an inventor, when the validity and scope of his patent has been assailed or he wishes to sell his patent, has found that he has little or nothing of value for the reason that his application for a patent was not properly prepared and prosecuted while pending in the Patent Office.

CHAPTER IV.

WHAT IS PATENTABLE.

In the United States patents are issued for a machine, an article of manufacture, an art or process, a composition of matter, a design.

Machine.—The Standard Dictionary defines a machine as any combination of inanimate mechanism for utilizing or applying power. This broad view is the one adopted by the Patent Office and the courts.

Any new and useful machine is patentable. As to utility, this may be nominal ; that is to say, if the machine is at all useful and is new, it is patentable, but the machine must be new or an improvement on existing machines. As to novelty, this does not usually consist in a wholly new machine, for it is very seldom that a machine is made with new parts. In fact, it is doubtful if one is ever made in which all the parts are new.

Usually a machine consists of a combination of elements old in themselves, but combined in a new way so as to accomplish a new result or to accomplish an old result in a new or better way. The one essential is that there must be some new operative change in the machine.

If the difference in construction between the new machine and the old is slight but the difference in results is obvious, then there is invention and the novelty contained in the structure is patentable. But let the machine be ever so new, its parts or operative means and not its principle must be claimed, and while a skillful attorney will draw claims broad enough to cover all analogous structures and so secure the field to the inventor, still, within the meaning of the United States Patent Law, one cannot claim a principle, as a principle is too indefinite and intangible to come within the scope of the claim.

Let it be borne in mind in connection with this subject and those immediately following, that an invention is not necessarily a new creation, but the inventor may simply perceive a means or a way of bettering mechanical contrivances and accomplish the result by improved means which involve sufficient novelty to come within the scope of the term invention and to entitle him to a patent. That is, the improved means involve something more than mere mechanical skill.

A Manufacture.—A manufacture is anything made by industrial art or processes or skill, whether it be made by hand or by machinery. Generally speaking, an article of manufacture, as contemplated by the patent law, comprises any vendible article of trade or commerce which is not a machine or a composition of matter; for example, a bag, a chair, or a shoe is an article of manufacture.

Composition of Matter.—Within the meaning of the patent law, a composition of matter is a combination of two or more substances making a substance which has some useful function. A well-known example of this kind is an explosive, a plating compound, a polishing substance, a substance for removing hair from hides or grease from leather, et cetera.

In applying for a patent as stated in another chapter, the applicant must specify the ingredients, the proportions in which they are mingled and the manner of combining them, whether chemically or otherwise, so that any person skilled in the art can, from his description, make the new composition.

Improvements.—An improvement, as its name indicates, is usually an advance made in an art or the construction of a machine which improves upon one already existing. This is the character of most inventions. Sometimes a person will invent or discover something entirely new, like Morse's invention of the telegraph, or Bell's invention of the telephone, and a multitude of inventors will follow after and improve upon the original device; such improvements, if they are really improvements or if they materially affect the function of the device, are patentable.

Art or Process.—An art or process, within the meaning of the patent law, is a method of reaching or accomplishing a certain result as distinct from the result itself or from the mechanism or means for accomplishing the process.

An art is the most comprehensive of inventions as it may include practically or may really cover both the method or process and the instrumentalities used in the operation. Some means must be described for carrying the process into effect, but in order that the process and the instrumentalities or apparatus may be included in one patent, they should be so dependent, one on the other, that they cannot practically be separated.

Broadly and generally an art or process is a new operative means for accomplishing a certain result.   A patent for an art is usually the broadest kind of a patent.   For example, if a process comprises three distinct steps in the treatment of a certain subject-matter, the claim will cover those three steps and it will not matter whether the steps are performed by hand, by machinery or in what way they are performed.   The mere fact that they are performed by an unauthorized party will constitute an infringement of the claim. It should be clearly understood that the art or process is entirely distinct from any mechanism employed in carrying the art or process into effect, although if the mechanism and the process are dependent one on the other, both may be included in the same patent and both the process and the apparatus or mechanism covered by independent claims. But though an art comprises so much, it must be capable of producing tangible, physical results, or else it is too indefinite to come within the purview of the patent law.

This subject of what can be included in a claim for an art and what should be included is one requiring the utmost skill and discrimination. The claims should include only what can be rightly claimed under a patent for an art or process. They must not be so broad as to claim an inoperative art, they should not be so narrow as to limit the patentee too much in practicing the art or process, but there are so many nice distinctions relating to this matter of claims that it must be left to the attorney, who will judge by the circumstances of each individual case.

Designs.—The statute relating to designs reads: "Any person who, by his own industry, genius, efforts and expense, has invented and produced any new and original design for a manufacture, bust, statue, alto-relievo, or bas-relief; any new and original design for the printing of woolen, silk, cotton or other fabrics; any new and original impression, ornament, patent (pattern), print, or picture to be printed, painted, cast or otherwise placed on or worked into any article of manufacture; or any new, useful and original shape or configuration for any article of manufacture, the same not having been known or used by others before his invention or production thereof, or patented or described in any printed publication, may, upon the payment of the fee prescribed, and other due proceedings had the same as in cases of inventions and discoveries, obtain a patent therefor."

Design patents are issued for three and a half, seven and fourteen years and the applicant must elect, when he files his application, for which term he will have his patent issue.

Inventors frequently have an idea that they can procure design patents cheaply and that they will cover the same ground as a patent for a structure. This is not usually so. Design patents relate exclusively to the shape, outline or configuration of any figure, article, print, fabric and the like. The claim being merely to the configuration, then, it is obvious that any radical departure from such configuration will not be an infringement of a patent for a design. Generally speaking, the rule is the same as in regard to an infringement of a trade-mark, that is: Will an ordinary purchaser be deceived and mistake one design for the other?

While the claim of a design patent relates exclusively to the configuration or shape, outline or ornamentation, still there are cases where articles of manufacture and even mechanical articles are valuable chiefly because of their peculiar shape and in such case a design patent is the proper means of protecting them, because, as a rule, a patent for the structure cannot be obtained. Any new configuration can be covered by a design patent even though it may not be purely ornamental. For example, a man may have a machine frame of a new shape, which shape is advantageous or which increases the value of the machine, and such a frame may be covered by a design patent. Perhaps it may be necessary to cover the frame of a cultivator, the body of a carriage, ornamental designs of fabrics, such as carpets, laces, and a thousand and one other articles, and so long as the form or configuration is essential, the design patent is usually the proper and often the only means of protection.

CHAPTER V.

CAVEATS.

Caveats, while still allowed by the statute and filed to a certain extent in the Patent Office, are gradually falling into disuse. A caveat is really of little value to the inventor. When a person has an invention, more or less complete, he can file a written description of the invention, together with a drawing, if this can be done, in the Patent Office by paying the prescribed fee. The caveat will be in force for a year, and can be renewed from year to year by renewing the fee.

Caveats can be filed by citizens of the United States only, though it has been recommended to Congress that this privilege be extended to foreigners.

The caveat does not protect the inventor and does not give him any right, as, for example, the right conferred by the issue of a patent, but it merely entitles him to notice in case some other party applies for a patent for substantially the same thing while his caveat is in force.

As the caveator has not an exclusive right, the only advantage from such a notice is that he can himself file an application for a patent, and thereupon he will be declared in interference with the other party who has filed an application for a patent on a similar article and has filed similar claims. This advantage is of a doubtful character, and the caveat really amounts only to evidence ; that is to say, it is good evidence that at the time of filing his caveat he had an invention in as complete a condition as his caveat papers show. If he then files his application, and claims the same matter claimed by the other applicant he has an interference suit on hand. The subject of interferences will be treated hereafter, but as it is mentioned incidentally here, it may be well to say that an interference, so styled, is a contest between applicants who claim a patent on the same invention, and as the patent obviously cannot be issued to both, this contest has to be first settled, and the patent will be issued to the one who proves to be the first inventor. This matter of interferences is a very important one, the practice concerning which is intricate and not thoroughly well settled in every particular, but the subject will be treated in a separate chapter.

Concerning the subject in hand, to wit, caveats, it will be clearly seen from the few remarks relating to them that a caveat is merely evidence, and that the inventor would be as well off if he had merely made a drawing of his invention and had it witnessed by reputable persons who could make oath that they had seen it at a certain time.

If instead of filing his caveat, he had made his invention sufficiently complete to enable him to show an operative device, and had filed his application for a patent, he would have been much better off, because the first applicant has a decided advantage in an interference case, and the burden of proof is on the second applicant, who must show by strong proof that he was the first to conceive and was using reasonable diligence to reduce the invention to practice.

There are cases where it may be advisable to file a caveat. Such a case might be, perhaps, where a party has in mind a complex invention, partially completed, and only knows in a general way how he will work out the details and complete the invention. In such case it may be to his advantage to file a caveat for the invention as far as completed, which would be evidence itself of having made the invention at the date of filing the caveat and to the extent disclosed therein.

The practice of filing caveats is not recommended to the average inventor; first, because of the inadequate protection of the caveat, and, second, because of the expense—that is to say, while a caveat is not in itself very expensive, still, if properly filed, it requires the preparation of drawings and specifications by an expert, or, at least, a specification, and the cost of this, taken in connection with the Government fee of ten dollars, is something to the average inventor, who is not supposed to be very wealthy.

The fact that caveats are of little value is largely owing to the further fact that while the caveator is entitled to notice in case another files an application for a similar thing, still the Patent Office is not bound to give him such notice, and he has no remedy for the neglect of the Office to notify him.

Moreover, as the caveat fees do not apply on the patent fees when the patent application is made, the cost of the caveat seems to be in nearly every instance so much time and money wasted. The better practice is for the inventor to complete his invention at as early a date as possible and make application for letters patent.

CHAPTER VI.

WHO MAY OBTAIN A PATENT.

Section 4886 of the Revised Statutes says that "Any person * * * may * * * obtain a patent." The words "any person" have been construed to mean a man or woman, whether the woman be married or single; and a minor, male or female, as well as any person embraced in these classes and also an alien. It has been further held that any number of persons whose combined efforts resulted in bringing forth an invention could make application as joint inventors and the patent would issue to them.

In many foreign countries, the first to introduce an invention can obtain a patent, as, for instance, in Great Britain. But in the United States a valid patent can only issue on the application of the real inventor or inventors, who must make oath to the invention and if the application and oath are made by one who is not the real inventor and the patent afterward issues to him the patent will be held invalid if the facts in the case are proved. Further, the invention and patent to issue may be owned by some person or corporation other than the inventor, but, notwithstanding this fact, the inventor must make the application.

If the invention has been assigned and the inventor refuses to make the application when under obligation to do so, the owner can apply to the proper court and get an injunction restraining the inventor from disposing of the invention and can compel him to make the application. If there is no assignment, but only an agreement to assign, he can be compelled to execute an assignment of the invention to the proper party.

Some women are prolific inventors and many of their inventions are and have been of great value. This is true to such an extent that the Patent Office of the United States has published and has for sale pamphlets styled "Women Inventors" and the mere list makes quite a respectable volume.

In connection with this subject of women inventors, it may be of interest to note that one of the first inventions of which there is any record was made by a woman. To quote from a recent reported lecture of Rev. Dr. N. D. Hillis : " A thousand years ago the race dwelt to the east of the River Jordan. Men came to little caves and these little caves had little doors, and these doors were hung on hinges.

Years before that, a young girl, with a bloom on her cheek, lived with her parents in one of these caves, and one Sunday night came a young man from over the hills to see the girl's father, the first time. The next day she said: 'Now, father, why couldn't this family have two caves, one for home folks and one for company?' And the father said yes, and the next Sunday night the young man came over to see the girl's mother, perhaps, and after that the young girl said: 'Father, we ought to have a door between the two caves so that it can be shut.' And she wanted a door hinge, that the door might be closed at will between the two caves. They had never seen hinges, so she set her wits to work to invent a door hinge, and she made one of the first inventions man ever saw. No, she didn't invent it at all; she copied it. Here is the model of all the hinges in the world, the hinge in the elbow. The other day a man was digging in the sand to the east of the Jordan and came upon a tablet on which was the image of a young girl; in her left hand she has a little chisel and in her right hand she has a large hammer. In front of her is a door hinge, and over at this end is the outline of a little elbow. That has been preserved for five thousand years to tell us how this young girl invented the first tool that the world ever saw."

Joint Inventors.—Where an invention is the joint product of two or more minds working together application must be made by all the parties who contribute to the invention. It is not necessary to make joint inventors, that one should produce or invent a distinct part of a machine, a second, another distinct part, because, if this is the case and the distinct part mentioned constitutes an operative device, each must apply for a patent on his own invention, but if there is a joint contribution, that is to say, if one brings, for instance, the general idea, another contributes certain improved details and they thus work together, one suggesting and another improving, they must join as applicants or else the patent, when it issues, will be invalid.

It must be understood, however, that there is a distinction between invention and skill.

It is very unusual for an inventor who is not a mechanic to employ a skilled workman to carry his ideas into effect, but this does not make the mechanic an inventor. Not infrequently, a skillful mechanic who is in the employ of a great inventor and really does good work, will make the statement, which is given more or less credence, that he is the real inventor of such and such a thing. The inventor is not supposed to be capable of doing all things and he has an undoubted right to obtain the best skill obtainable to carry out his ideas; that is to say, he will usually get a skilled draughtsman to give his ideas good mechanical shape, and his machine, if it be a machine, suitable design. He will also get good mechanics to make the parts, assemble them and make such changes as may suggest themselves to their practical minds, but notwithstanding the fact that the work may be very skillful, still, so long as the inventor brings the ideas to the workman, his invention is not thereby impaired and he has a perfect right to apply for and obtain a valid patent.

If, however, the workman by his skill contributes to the real substance of the invention, as a whole or only as to part, and he carries into effect ideas not thought of by his employer, he must take out the patent himself in the one case or be joined as an applicant in the other. In this case not much skill is exercised, but invention. If the patent is to issue to one other than the inventor or to the inventor and some other person not an inventor, this must be effected by a proper assignment.

Persons employed to do skilled work have a right on their own time to carry into effect independent inventions, but their ideas must be entirely independent from those of their employers, and must be such improvements or must embrace such changes of mechanism as would not suggest themselves to an ordinary skilled mechanic. If the suggestions of the mechanic really constitute the complete machine, and the one posing as the inventor merely suggests that he would like to do certain things, without specifying means by which the result is obtained, and the mechanic's ideas are shaped and made to accomplish the desired result, then the mechanic is the inventor.

Employer and Employee.—One has the right to hire a person for the purpose of inventing, but in such case the employee must sign any application and the patent to issue legally to the employer must be duly assigned to him. If the employee, while in the general employ of the employer, makes an invention on his own time and with his own materials, he has the legal right to the invention and his employer cannot interfere with this invention. If, on the other hand, the employee has made an invention on his employer's time and has used his employer's materials, in such case, the employee is still the inventor, but the employer has an implied license, not transferable, which a court of competent jurisdiction will enforce and which will give the employer the right to the use of the invention in his business.

This implied license, as above remarked, is not transferable and if the employer is a corporation, the license is extinguished by the dissolution of the corporation. It will be seen, then, that the employee, in such a case, has a perfect right to make any use of the invention he sees fit. He can sell it, lease it or do any act that any inventor and patent owner might do, but he cannot deprive his employer of the rights of a licensee.

A Deceased Inventor.—If a person makes an invention and dies before making an application for a patent, or before the application is completed, the application can be made or prosecuted by the executor or administrator. If the deceased leaves no will, the right to patent will go to his legal representatives.

Patent Office Employees.—Persons in the employ of the Patent Office are barred from procuring patents while in such employment, although an employee may properly obtain a patent after he has left the Patent Office.

CHAPTER VII.

CONCERNING PATENTABILITY.

What Constitutes Invention.—The statute requires that in order to obtain a patent one must invent or discover something new and useful. The popular definition of the word invention is the contriving or bringing out of something which did not before exist ; but the statute requires more than this.

Any fairly resourceful mechanic, such as a machinist, a carpenter, or other artisan, is capable of creating something which did not before exist, because the exigencies of his work require it. One will scarcely find two building exactly alike. Two machinists will hardly do their work in the same way in bringing forth a well-known machine, and so on through the whole realm of mechanics.

The workman who has had considerable experience has sufficient skill to enable him to meet the ordinary requirements of his trade and to depart from existing models to a certain extent, but he is not called upon to really create or invent anything ; that is to say, he is not required to use his inventive faculty to put together things in such essentially new ways as to accomplish different results from those heretofore obtained, or to combine things so that they will have functional differences from things already combined. The real distinction between invention and mere skill is that one is produced by original thought, while mere skill utilizes the discoveries of others, either by imitation or by employing good judgment in selecting and combining them, or in applying them to practical results.

If a person uses his inventive faculty, and really gives to the world or to the public something new and useful, something in which the result obtained is real and tangible, he has done the public a service and has given a *quid pro quo* for the patent which will issue to him, but it is not the intent of the statute to offer a reward merely for skill, no matter how great its order.

It frequently happens that a person will discover a new use for an old thing, but this does not amount to invention, even though the result is very important, unless some change is required to adapt the thing to the new use. This is what the Patent Office and courts term Double Use, and it follows that if a person merely discovers that a tool previously used for certain purposes can be used to advantage for another purpose he has not made an invention, but simply has enlarged the use of a well-known object. Moreover, the invention which he makes must be one that is not obvious; that is to say, that is not the result of mere skill.

Issued patents are open to the public and any subsequent inventor is presumed to know of their existence, even though they may cover subject-matter which has never been put in practical use. The issued patents may cover a machine which has never been made, still the fact that the drawings and description of the invention are on file at the Patent Office is a notice to any subsequent inventor, as if the machine had really been built. If, then, the new invention is one that would readily suggest itself to a person of ordinary skill, after a perusal and examination of existing patents, then the man cannot be held to have invented anything, but he has merely followed out obvious suggestions.

The application of an old process to manufacture an article to which it had never before been applied is not a patentable invention, nor does the application of old machinery to a new use involve invention. Aggregations of well-known things do not form inventions within the meaning of the statute. If stove hooks, socket wrenches and screw-drivers be old, as we know they are, then if one provides a single tool having a hook at one end and the well-known form of a screw-driver of the ordinary kind at the other end and a wrench socket at some convenient place on the handle, he has invented nothing, even though such combination never before existed. He has merely aggregated and collected a series of well-known things. This illustration serves well to show the difference between the combination of old parts to produce a new result and an aggregation. It will be noticed in the aggregation referred to that the stove hook serves simply as a stove hook, the screw-driver as a screw-driver, and the wrench as a wrench. Neither part coöperates with the other to accomplish any result, but in a legitimate combination, the several parts of the combination coact to produce a certain definite and tangible result.

It sometimes happens that one part may have an independent function, but it may also have a combined function with the other parts or elements of the combination. In such case a claim may be made for the part having the independent function, and another claim for the part in combination with the other elements, with which it coacts.

The examples given of what does not constitute invention will perhaps be as good a guide as any as to what does constitute invention. It is a general principle that mere changes in the

size or form of a thing or the number of articles
composing a whole, or the degree of curvature
or other dimension, as the shape of a dye, does
not amount to invention.   Neither does merely
the substituting of one material for another.
Supposing wooden door knobs to be common,
it would not involve invention to substitute
porcelain for wood, even though porcelain had
never before been used for a door knob.

If the porcelain were a new composition, then
the inventor could cover it as a composition of
matter without regard to the use to which the
composition may be put.

Discovery and Invention Defined.—
Within the meaning of the Patent Law, a patentable invention must possess a certain amount of
utility, and must have some new feature, or produce some new result, not obvious from any
source of information, which makes the invention new or else gives it a function or use not
heretofore existing.

A patentable discovery consists in first find-
ing some principle or law of nature within the
range of patentable subject-matter and reducing
the same to practice.

Tests of Patentable Novelty.—As previously stated, probably one of the best tests of
patentable novelty is this: Is there anything
existing in the art which would naturally suggest to a skilled mechanic the alleged inven-

tion ? If there is an essential change in function, it would be held as generally true that there is invention. If the change of function is not essentially obvious, but the result achieved is important ; for example, if it is a machine and it makes a given article at a less cost or makes better articles or has essentially better results in any way, this fact of betterment is significant and it is almost conclusive evidence that the new matter involves invention. Or as one judge says : " While it is true that the utility of a machine, instrument or contrivance, as shown by the general public demand for it when made known, is not conclusive evidence of novelty and invention, it is, nevertheless, highly persuasive in that direction, and in the absence of pretty conclusive evidence to the contrary, will generally exercise controlling influence."

A decided advance in the art or in the result, even though accomplished by means quite similar to something heretofore existing, is ordinarily good evidence of invention.

Some Inventions not Patentable.—It is possible for one to use great ingenuity and invent something which has never before existed and which is useful and valuable, and yet may not come within the purview of the patent law. It is not unusual for a person to invent a certain advertising scheme which is an excellent thing, which enables the advertiser to do better advertising than has been done before and yet the scheme is not patentable because it cannot be put in such tangible shape as to come under the head of a machine, a manufacture, an art or process, or a composition of matter.

Many patents are taken out on advertising devices and some of them are very remunerative. But there is a difference between a device and a scheme or method of advertising. The latter is not patentable. It is a mental process pure and simple. And so one may have a new business method or scheme which is ever so ingenious, which is valuable, but which cannot be put in such shape as to be patentable or does not come under the classification of patentable subjects. Likewise, one may have a method of bookkeeping, which would render the work easier or more accurate, but this comes under the head of unpatentable and intangible things. If the invention is in a book and its peculiar arrangement, the book may be patented, but not the method. When one has evolved an idea along these lines all he can do is to get what he can out of it, before other people discover the scheme, but he cannot invoke the protection of the law.

Inventions which are against public policy are not patentable, and while a hard and fast rule cannot be laid down as to what inventions are against public policy, still the inventor usually knows whether or not such is the case. It has been held that slot machines used merely for gambling purposes are inventions of this class and as such are not patentable.

CHAPTER VIII.

PRIOR USE—PUBLIC USE—EXPERIMENTS.

The statute provides that an invention to be patentable must not be known or used before applicant's invention or discovery thereof, or patented or described in any printed publication in any country before his discovery or invention thereof, or more than two years prior to the application, or in public use or on sale in the United States for more than two years prior to the application for the patent, unless the same is proved to have been abandoned. The statute is not quoted, but only its substance given. It is provided further by statute that it is a good defense for an action of infringement of the patent to show that the patentee was not the original and first inventor or discoverer of the invention patented. Such prior use means use by another than the inventor, the knowledge and use occurring prior to the patentee's invention.

The use or knowledge of the invention abroad will not affect the patent here, providing it had not been patented or described in a printed publication anywhere before its invention in this country. A foreign patent or publication to anticipate an application or invalidate a patent must disclose substantially the same invention, and must have been made public before the person who obtained the American patent made the invention.

In attempting to take advantage of the defense of prior use, it very often happens that the defense will make a reference to some alleged prior invention containing the substance of the thing patented, when, as a matter of fact, the prior invention was in the nature of an experiment, and was abandoned. These abandoned experiments are not of such a nature, within the meaning of the statute, as will invalidate a patent. The burden of proof, when prior use is set up, is on the defendant, and it takes strong proof to establish such a defense. If a prior patent has been issued for the same invention as that disclosed by a later patentee, even though the claims are not identical, the prior patent will invalidate the subsequent one unless the subsequent patentee can show that he was really the first inventor.

Two Years' Public Use.—The inventor may have made, used or sold his invention in the

United States to a great extent and for profit, but this will not debar him from obtaining a valid patent if such use has not been continued for more than two years prior to his application for patent, and he can honestly make oath to the fact that the invention has not been in public use or on sale in the United States for more than two years, even though he may have been at work on the invention for many years, so long as his work was of an experimental nature. If, however, the inventor has made or sold or used the article publicly for more than two years prior to his application, and this fact can be proved, he cannot obtain a valid patent, as a court will construe such long public use as an abandonment of the invention to the public. Nor can any inventor obtain a valid patent if he allows it to be used by persons generally, either with or without compensation.

Experiments.—In connection with this subject of public use and prior use is the matter of experiments. A public experiment is never public use within the meaning of the statute, so long as the experiment is bona fide and is for the purpose of testing the qualities of the invention. If the inventor uses the invention for profit, and not for experimental purposes, that is public use, though in some instances, if the profit was incidental, or it was necessary to result in profit to show the inventor how to perfect his invention, and was so used, such would not amount to public use. The experiment may have been used in public every day for several years, and have been known to hundreds of persons, and yet not be a public use.

BOOK II.

CHAPTER I.

THE APPLICATION.

The matter of making the application for a patent is one of supreme importance to the inventor. It may be that he has an invention such as will never come to him again, that is of more importance than any other to him, and it is therefore absolutely necessary that all his rights be properly safeguarded. Moreover, if the invention is worthy of patenting at all, it . is worthy of patenting well.

Patent Office practice is intricate and peculiar and only those persons who are naturally qualified for this work and have added to their natural qualifications by study and experience are competent to attend to it. Many people who have not had experience in this line think that if they get a patent under the seal of the Patent Office this covers everything, and that a patent is a patent. Nothing could be farther from the truth.

A patent application embodies a petition, specification, drawings and oath ; the details of which will be hereinafter referred to. The patent specification is one of the most difficult instruments to draw properly and if an incompetent or negligent person has charge of the application and prosecution of the case, he is likely to let the patent go to issue with claims which will not protect the inventor when reasonable diligence and skill would have given him adequate protection. Every examiner in the Patent Office is often exasperated to see the manner in which the inventor's interests are sacrificed. In an important invention it often happens that the subject-matter is essentially new, but that the claims have been drawn so that in terms they are not allowable, whereas by proper amendment they would be so.

In such a case the examiner must reject the claims. An incompetent or unscrupulous attorney who is anxious to get a quick allowance will sometimes cancel a lot of rejected claims and permit the patent to go to issue upon claims which are wholly insufficient. The examiner at the Patent Office is not at fault and is practically powerless to help the inventor. The patentee may probably know nothing of the matter until in the exploitation of his invention he attempts to sell it or to get money to properly develop and to market it. The prospective investor, being a man of the world and having experience, may be pleased with the invention, but he will almost invariably refer the matter to his counsel to see if the patent is valid and the invention properly covered by the claims. His counsel will, on investigation, see that the patent is wholly inadequate and must so report; consequently the patentee is unable to interest capital and sees a fortune slip away from him, and may have the further mortification of seeing the same person whom he has tried to interest making a fortune out of an invention so similar to his that it would have been a palpable infringement if his patent had been properly obtained.

Business men usually recognize the rule that cheap help is the most expensive ; that is to say, it is not profitable to put a professional man to digging a ditch, but it is profitable to employ the best ditch-digger if a ditch is to be dug and pay him reasonably, rather than have a poorer man for less money. Good counsel is always the cheapest in the end and the difference in cost is not so very great. The majority of patent agents who advertise widely and work cheaply are either incompetent or else work for such small fees that they cannot afford to do proper work.

The Patent Office advises an applicant unless familiar with such matters '' to employ a competent attorney as the value of patents depends largely upon the skillful preparation of the specification and claims." There is no excuse for an inventor if he suffers through poor legal advice and skill. While there are, no doubt, many incompetent attorneys in the country, still there are in most large cities competent and reliable attorneys who work for reasonable fees and who are thoroughly competent to take care of their clients' interests. The inventor will have to pay such person, perhaps, from ten to twenty-five dollars more for the prosecution of an ordinary Patent Office case, but by the expenditure of this small additional sum, he may save himself a fortune in the end. It is of the utmost importance for any person, inventor or otherwise, having any patent work of any nature whatsoever to employ good counsel.

Some of the most important statutes relating to patents are in their essential parts as follows :

SECTION 4884.—Every patent shall contain a short title or description of the invention or discovery, correctly indicating its nature and design, and a grant to the patentee, his heirs or assigns, for a term of seventeen years, of the exclusive right to make, use and vend the invention or discovery throughout the United States and the Territories thereof, referring to the specification for the particulars thereof. A copy of the specification and drawings shall be annexed to the patent and be a part thereof.

SECTION 4886.—Any person who has invented or discovered any new and useful art, machine, manufacture or composition of matter, or any new and useful improvements thereof, not known or used by others in this country before his invention or discovery thereof, and not patented or described in any printed publication in this or any foreign country, before his invention or discovery thereof, or more than two years prior to his application, and not in public use or on sale in this country for more than two years prior to his application, unless the same is proved to have been abandoned, may, upon payment of the fees required by law, and other due proceeding had, obtain a patent therefor.

SECTION 4887.—No person otherwise entitled thereto shall be debarred from receiving a patent for his invention or discovery, nor shall any patent be declared invalid, by reason of its having been first patented or caused to be patented by the inventor or his legal representatives or assigns in a foreign country, unless the application for said foreign patent was filed more than seven months prior to the filing of the application in this country, in which case no patent shall be granted in this country.

SECTION 4889.—When the nature of the case

admits of drawings, the applicant shall furnish one copy, signed by the inventor or his attorney in fact, and attested by two witnesses, which shall be filed in the Patent Office ; and a copy of the drawing, to be furnished to the Patent Office, shall be attached to the patent as a part of the specification.

SECTION 4890.—When the invention or discovery is of a composition of matter, the applicant, if required by the Commissioner, shall furnish specimens of ingredients and of the composition, sufficient in quantity for the purpose of experiment.

SECTION 4892.—The applicant shall make oath that he does verily believe himself to be the original and first inventor or discoverer of the art, machine, manufacture, composition or improvement for which he solicits a patent ; that he does not know and does not believe that the same was ever before known or used ; and shall state of what country he is a citizen. Such oath may be made before any person within the United States authorized by law to administer oaths, or when the applicant resides in a foreign country, before any minister, chargé d'affaires, consul, commercial agent, holding commission under the Government of the United States, or before any notary public of the foreign country in which the applicant may be.

SECTION 4893.—On the filing of any such application and the payment of the fees required by law, the Commissioner of Patents shall cause an examination to be made of the alleged new invention or discovery ; and if on such examination it shall appear that the claimant is justly entitled to a patent under the law, and that the same is sufficiently useful and important, the Commissioner shall issue a patent therefor.

SECTION 4894.—All applications for patents shall be completed and prepared for examination within one year after the filing of the application, and in default thereof, or upon failure of the applicant to prosecute the same within one year after any action therein, of which notice shall have been given to the applicant, they shall be regarded as abandoned by the parties thereto, unless it be shown to the satisfaction of the Commissioner of Patents that such delay was unavoidable.

As will be seen by reference to the statutes, it is necessary in a complete application to have drawings of the invention where the latter admits of drawings, a petition, a specification which will describe the invention so accurately that it can be carried into effect by those skilled in the art from the specification, and drawings if any, and an oath. It will be assumed that the inventor uses good judgment and employs good counsel to prepare and prosecute his case. If he can see his attorney, the latter will get all the information he requires, but whether he sees him or not, and particularly if the matter is attended to by correspondence, the inventor should be careful to conceal nothing whatever from his attorney, but to go into the utmost detail and specify every matter which really concerns the invention. That is to say, it is not necessary for him to refer in great detail to immaterial matters, but let him thoroughly describe his invention, and particularly the advantages of the new construction. Let him discriminate between the new and the old, and point out with great particularity all the differences between the old and the new, so far as he can, as this materially assists in properly preparing the application papers.

The specifications and claims, by their scope, render the patent broad and sufficient, or narrow and incomplete. A good attorney will bring out in the specification all the essential matters, describe the new functions accurately and well, and so lead up to claims which he will draw in such a way as to thoroughly cover the invention. The attorney will prepare the necessary petition and oath, and after the papers are executed will file them in the Patent Office, paying the first Government fee of fifteen dollars.

Filing of Application.—The application is examined in its order of filing—that is to say, in the order in which it is received in relation to other applications. The applications are filed in certain divisions of the Patent Office according to the nature of the invention, and they are taken up by the examiner in charge of the division, in their regular order, and this order is not departed from except in the following instances :

(1) Applications wherein the inventions are deemed of peculiar importance to some branch of the public service, and when for that reason the head of some department of the Government requests immediate action and the Commissioner so orders ; but in such case it shall be the duty of such head of a department to be represented before the Commissioner in order to prevent the improper issue of the patent.

(2) Applications for reissues.

(3) Applications which appear to interfere with other applications previously considered, and found to be allowable, or which it is demanded shall be placed in interference with an unexpired patent or patents.

Examination of Application.—When the case is reached by the examiner, he goes carefully over the specification, drawings and other papers, and if there are any technical objections, such as typographical errors, irregular claims or insufficient or improper drawings, he calls the attention of the applicant or his attorney to this matter, and usually at the same time he passes on the merits of the claims. In construing the claims and either allowing or rejecting them, he considers the state of the art to which the application appertains and compares the case with existing United States patents in the art, goes over such foreign patents as are obtainable, and even examines public documents, catalogues and technical books, because it must be borne in mind that the invention must be new in order for the patent to issue, and if claims pass to issue which can be construed to cover existing matter they are invalid. After making such an examination the Patent Office makes a report to the attorney of record, if there is one, and allows such claims as are not found objectionable and rejects the others.

It is here that the skillful attorney is of especial use to the inventor. The examiners at the Patent Office are, as a rule, conscientious, careful and well skilled in their art, but they are human, and therefore not infallible. Not infrequently they will reject a claim which is really allowable, and if the attorney is sufficiently skillful and discriminating, he can usually present the case to the examiner so that he will see that the claim or claims may be allowed, and the other claims the attorney will cancel or amend, as the necessities of the case require, until no claims are left in dispute but all remaining are allowable, after which the official notice of allowance is issued.

The applicant then has six months in which to pay the final Government fee of twenty dollars, and when this is paid the patent issues to him. The fee can be paid at once or at any time during the six months. If the attorney and the examiner cannot agree as to the allowability of the claims, the applicant can appeal. It sometimes happens that two or more applications for the same subject-matter, but by different inventors, will be pending in the Patent Office at the same time, or that a pending application will be rejected on a patent issued within two years of the filing of the application, and that the applicant can show that he has made his invention before the filing of the prior patent. In either of these cases the parties are declared in interference. This matter of appeals and interferences will be considered in separate chapters.

CHAPTER II.

APPEALS.

The rules of practice relating to the prosecution of patent applications and of appeals in patent cases are extremely liberal to the inventor and would-be patentee and give him the opportunity to obtain his rights without undue expense. The claims are not left to be decided by the individual who, though ever so honest, might do the inventor great injustice. But if the applicant for a patent is dissatisfied with the ruling of an examiner in the Patent Office who rejects a claim or claims, he has the right to appeal from the decision of the primary examiner to the Board of Examiners-in-Chief; from the Examiners-in-Chief to the Commissioner of Patents, and from the Commissioner of Patents to the Court of Appeals for the District of Columbia, so that there is ample means for him to obtain his deserts.

The practice of the primary examiner in making the examination and passing on the patentability of claims has been already gone into in the preceding chapter, but it sometimes happens that the examiner and the applicant or his attorney cannot agree on claims which are thought to be allowable. In that case the applicant has a right, as above stated, to appeal to the Board of Examiners-in-Chief. This Board is composed of three skilled persons, competent to pass on the scope and patentability of claims, and to them the appeal is taken in the first instance. For such an appeal there is a Government fee of ten dollars, but if there are questions which do not involve the patentability of the claim or affect the merits of the invention, for instance, as to the question of division or whether an amendment shall be entered and considered, he can petition the Commissioner without fee, and have the question considered and decided by him.

Before an appeal can be had to the Board of Examiners-in-Chief, the claim must have been twice presented and twice rejected. The hearing before the Board of Examiners-in-Chief is substantially the same as a hearing in court. The appellant files his reasons for appeal and usually a brief of the authorities and arguments on which he relies to maintain his appeal, and if he desires he can have an oral hearing before the Board. If the decision is adverse to him he can appeal to the Commissioner in person upon the payment of the fee of twenty dollars, and the Commissioner reviews the decision of the

Board and can either affirm or reverse it and, as before remarked, if he desires the applicant may appeal from the decision of the Commissioner to the Court of Appeals for the District of Columbia. In each case of appeal the appellant must state his reasons for the appeal; state wherein the person or persons making the decision appealed from erred, and should file a brief setting forth the grounds of appeal, authorities on which he relies, etc.

CHAPTER III.

An interference is a proceeding instituted for the purpose of determining the question of priority of invention between two or more parties claiming substantially the same patentable invention. The mere fact that one of the parties has already obtained a patent will not prevent an interference, for, although the Commissioner has no power to cancel a patent, he may grant another patent for the same invention to the person who proves to be the prior inventor. Interference according to practice in the Patent Office, will be declared in the following cases when all parties claim substantially the same patentable invention :

(1) Between two or more original applications containing conflicting claims.

(2) Between the original application and an unexpired patent containing conflicting claims, when the applicant, having been rejected on the patent, shall file an affidavit that he made the invention before the patentee's application was filed.

(3) Between an original application and an application for the reissue of a patent granted during the pendency of such original application.

(4) Between the original application and a reissue application, when the original applicant shall file an affidavit showing that he made the invention before the patentee's original application was filed.

(5) Between two or more applications for the reissue of patents granted on applications pending at the same time.

(6) Between two or more applications for the reissue of patents granted on applications not pending at the same time, when the applicant for reissue of the later patent shall file an affidavit showing that he made the invention before the application was filled on which the earlier patent was granted.

(7) Between a reissue application and an unexpired patent, if the original applications were pending at the same time, and the reissue applicant shall file an affidavit showing that he made the invention before the original application of the other patentee was filed.

(8) Between an application for reissue of a later unexpired patent and an earlier unexpired patent, granted before the original application of the later patent was filed if the reissue applicant shall file an affidavit showing that he

made the invention before the original application for the earlier patent was filed.

(9) An interference will not be declared between an original application filed subsequently to December 31, 1897, and a patent issued more than two years prior to the date of filing such application, or an application for a reissue of such patent.

Before the declaration of interference all preliminary questions must be settled by the primary examiner and the issue must be clearly defined. The invention which is to form the subject of the controversy must be decided to be patentable and the claims of the respective parties must be put in such condition that they will not require alteration after the interference shall have been finally decided unless the testimony adduced on the trial shall necessitate or justify it. It is not proposed here to go into the technicalities of an interference proceeding. Such matters are uninteresting reading for one who is not specially interested in an interference case and moreover the subject is comprehensive enough to require a volume to treat it with accuracy and detail. Further, no sane person, unless he be a patent attorney or familiar with interference proceedings, would think of prosecuting an interference case himself, but would get some competent lawyer to represent him. The practice in interference proceedings is intricate and requires the services of a lawyer of good skill and experience. The proceedings are in the nature of a contest in equity and the parties to the interference are required under proper rules to take testimony to prove when they conceived the invention, when they made drawings, if any, when they made a model, if any, what they have done in the way of making the complete invention and matters tending to show when the invention was conceived and completed, and generally such matters as will tend to prove who is the first inventor, which is the real question in issue.

Importance of an Early Application.— If several applications are pending for the same subject-matter at the same time, the parties will be in interference, but the one who first files his application is the senior party, so called, and it will take strong evidence on the part of others to overcome the probability that he, the first applicant, is the first inventor. Moreover, the one who first reduces an invention to practice takes the important step and so important in fact that if he has conceived the invention subsequent to the conception of another, but is the first to reduce the invention to practice and the other neglects unreasonably to reduce to practice, then the proceedings will be decided in favor of the one who has used diligence in putting his invention into practical shape. Now it

is held that a complete application for a patent
is a constructive reduction to practice, for, to
make a complete application, drawings and
a specification are required, which enable an
operative machine or device to be made, or a
process to be operatively carried out, or a composition to be put together in practical shape.
The necessity of an early filing of the application is therefore apparent.   Fortunately, interference proceedings are not so very common ;
on the other hand they are not unusual, and
therefore it is decidedly to the interest of inventors, for this and other reasons, to make an
application for a patent as soon as they can get
the matter into proper shape for such application.   It is believed that interference cases will
be rather less common than formerly because,
owing to a recent decision, the claims of different parties must be substantially alike in
order to constitute an interference.

Evidence of Invention.—No one can tell
whether or not he may be involved in an interference proceeding when he files his application
for a patent, and therefore a prudent inventor
will preserve evidence of his invention.   The
importance of making an early application by
the inventor has already been shown, but he
should also be able to furnish good evidence of
his invention, outside of his application.   When
he conceives the invention, it is well for him to

make a sketch of it if possible, sign and date the same, and have competent witnesses to the sketch. It is also well to make a complete working invention at as early a date as possible as he has then effected what is known as "reduction to practice." This is an important matter in case he gets into interference proceedings, for one may have an early conception of an invention and take no further steps to put it in practice and make it of value to the public. A later inventor may at once proceed to reduce his invention to practice; to build a machine, if it be a machine, or to complete the invention, whatever its character, and in such case the interference proceedings will, other things being equal, usually be decided in favor of the one who has used diligence in perfecting his invention even though he may not be the first to conceive. These matters of conception, reduction to practice, diligence, laches, etc., are each of great importance in interference proceedings and there are so many questions which arise and concern the relative importance of any one of them that no general statement can be made as to which is the most important, for the circumstances in each case affect this matter.

It can be laid down as a good rule, however, that the inventor to protect his interest should always first reduce the conception to tangible form by a drawing or description as soon as he can do so and have the matter witnessed. Second, that he should, as soon as possible, carry his conception into effect by building the structure, if it be a structure, or completing the invention, whatever its nature, and, third and most important, he should make his application for his patent just as soon as he can get the subject-matter for the application and can decide that the matter is worth patenting. If the testimony has been taken on both sides in an interference proceeding, the records of the testimony and the briefs of counsel are laid in the proper manner before the examiner of interferences and the case is usually argued by counsel, after which the examiner renders his decision on the case. The question of appeal is practically the same as already considered under appeals from the primary examiner, that is to say, either party can appeal from the examiner of interferences to the Board of Examiners-in-Chief, from the Board to the Commissioner in person, and from the Commissioner to the Court of Appeals for the District of Columbia.

CHAPTER IV.

DISCLAIMERS AND REISSUES.

A disclaimer is an amendment to a patent after issue, which disclaims some definite and specific part of the patent.

The object of a disclaimer is to avoid having a patent declared invalid in case of litigation, by reason of claiming more than the owner is entitled to as justly and truly his own, unless he has preserved the right to disclaim the surplus. This right may be lost by unreasonable neglect or delay to file a disclaimer in the Patent Office.

As stated in another chapter, each claim stands or falls by itself. In case of litigation a favorable decision will be rendered if only one claim is infringed. It is usual before bringing a suit to have an examination made into the validity of the patent, and in case counsel finds that one or more claims are probably invalid and he does not care to have them passed upon, he does not put these questionable claims in issue. This makes it unnecessary to file a disclaimer as to those claims, and principally on this account disclaimers are not very often used.

The sections of the statute relating to disclaimers are as follows:

SECTION 4917.—Whenever, through inadvertence, accident, or mistake, and without any fraudulent or deceptive intention, a patentee has claimed more than that of which he was the original or first inventor or discoverer, his patent shall be valid for all that part which is truly and justly his own, provided the same is a material or substantial part of the thing patented; and any such patentee, his heirs or assigns, whether of the whole or any sectional interest therein, may, on payment of the fee ($10.00) required by law, make disclaimer of such parts of the thing patented as he shall not choose to claim or to hold by virtue of the patent or assignment, stating therein the extent of his interest in such patent. Such disclaimer shall be in writing, attested by one or more witnesses, and recorded in the Patent Office; and it shall thereafter be considered as part of the original specification to the extent of the interest possessed by the claimant and by those claiming under him after the record thereof. But no such disclaimer shall affect any action pending at the time of its being filed, except so far as may relate to the question of unreasonable neglect or delay in filing it.

SECTION 4922.—Whenever, through inadvertence, accident, or mistake, and without any willful default or intent to defraud or mislead the public, a patentee has, in his specification, claimed to be the original and first inventor or discoverer of any material or substantial part of the thing patented, of which he was not the original and first inventor or discoverer, every such patentee, his executors, administrators and assigns, whether of the whole or any sectional interest in the patent, may maintain a suit at law or in equity, for the infringement of any part thereof, which was bona fide his own, if it is a material and substantial part of the thing patented, and definitely distinguishable from the parts claimed without right, notwithstanding the specifications may embrace more than that of which the patentee was the first inventor or discoverer. But in every such case in which a judgment or decree shall be rendered for the plaintiff, no costs shall be recovered unless the proper disclaimer has been entered at the Patent Office before the commencement of the suit. But no patentee shall be entitled to the benefits of this section if he has unreasonably neglected or delayed to enter a disclaimer.

These sections should be construed together, and the kind of disclaimer referred to differs from those which are embodied in the original or in reissue applications, as originally filed or subsequently amended, in which the disclaimant does not claim title to matter shown and described. It also differs from those made to avoid the continuance of an interference. These latter disclaimers must be signed by the applicant in person and must be duly witnessed, and require no fee.

It will be seen from the sections quoted that any owner of the patent or of any rights under the same may disclaim, but the disclaimer affects only the interests possessed "by the claimant (the one making the disclaimer) and by those claiming under him after the record thereof."

It appears, therefore, that if there are different owners of a patent, one may disclaim, thus affecting the rights of those claiming under him, while the others may have a different right in the patent because one party may think a disclaimer should be recorded while another may take a different view of the matter. A disclaimer should only be filed under the advice of competent counsel, and if filed at all it must be filed without unreasonable delay.

Reissues.—The reason for filing a reissue is to correct a patent which is inoperative or invalid. This occurs when the specification is defective or insufficient, or the patentee claims as his own invention or discovery more than he had a right to claim as new. If the error has arisen by inadvertence, accident or mistake, and without any fraudulent or deceptive intention, and the applicant otherwise complies with the law, a new patent will be issued to him.

Usually reissues are applied for merely to broaden the invention. It rarely occurs that a court will sustain a reissued patent where the invention is broadened, particularly when there has been unreasonable delay in making the application. If there has been a delay of two years or more this would, unless in exceptional cases, be held to be unreasonable. In fact, much less time than this has been held to be unreasonable, particularly where others have entered the field who would not have infringed the claims of the original patent.

In view of the fact that a reissued patent with broadened claims will generally be held to be invalid, it is of the utmost importance that the application for a patent in the first instance should be carefully prepared and the application skillfully prosecuted while pending in the Patent Office by those competent to attend to such matters. Otherwise, the inventor may find himself in possession of a patent which does not protect his invention and which is of little or no commercial value.

The principal statute relative to reissues is as follows:

SECTION 4916.—Whenever any patent is inoperative or invalid, by reason of a defective or insufficient specification, or by reason of the patentee claiming as his own invention or discovery more than he had a right to claim as new, if the error has arisen by inadvertence, accident or mistake, and without any fraudulent or deceptive intention, the Commissioner shall, on the surrender of such patent and the payment of the duty required by law, cause a new patent for the same invention, and in accordance with the corrected specification, to be issued to the patentee, or, in case of his death or of an assignment of the whole or any undivided part of the original patent, then to his executors, administrators, or assigns, for the unexpired part of the term of the original patent. Such surrender shall take effect upon the issue of the amended patent. The Commissioner may, in his discretion, cause several patents to be issued for distinct and separate parts of the thing patented upon demand of the applicant, and upon payment of the required fee for a reissue for each of such reissued letters patent. The specifications and claim in every such case shall be subject to revision and restriction in the same manner as original applications are. Every patent so reissued, together with the corrected specifications, shall have the same effect and operation in law, on the trial of all actions for causes thereafter arising, as if the same had been originally filed in such corrected form; but no new matter shall be introduced into the specification, nor in case of a machine patent shall the model or drawings be amended, except each by the other; but when there is neither model nor drawing, amendments may be made upon proof satisfactory to the Commissioner that such new matter or amendment was a part of the original invention, and was omitted from the specification by inadvertence, accident or mistake, as aforesaid.

The Patent Office requires when filing an application for a reissue that the applicant, besides the usual petition and oath, must file a statement on oath as follows:

(1) That applicant verily believes the original patent to be inoperative or invalid, and the reason why.

(2) When it is claimed that such patent is so inoperative or invalid "by reason of a defective or insufficient specification," particularly specifying such defects or insufficiencies.

(3) When it is claimed that such patent is inoperative or invalid "by reason of the patentee claiming as his own invention or discovery more than he had a right to claim as new," distinctly specifying the part or parts so alleged to have been improperly claimed as new.

(4) Particularly specifying the errors which

it is claimed constitute the inadvertence, accident or mistake relied upon, and how they arose or occurred.

(5) That said errors arose "without any fraudulent or deceptive intention" on the part of the applicant.

The government fee for reissue applications is thirty (30) dollars. Owing to the work involved to prosecute a reissue patent properly, even under the most favorable circumstances, it is necessary for the attorney usually to charge more for his work than for the preparation of original applications.

CHAPTER V.

ABANDONED, FORFEITED, REVIVED AND RENEWED APPLICATIONS.

An abandoned application is one which has not been completed and prepared for examination within one year after the filing of the petition, or which the applicant has failed to prosecute within one year after any action therein of which notice has been duly given, or which the applicant has expressly abandoned by filing in the Office a written declaration of abandonment signed by himself and assignee, if any, identifying his invention by title of invention, serial number and date of filing. Prosecution of the application to save it from abandonment must include such proper action as the condition of the case may require. The mere fact of offering an amendment will not save the case from abandonment, but the amendment must be in the nature of a proper response to the last official action on the case.

Revival of Abandoned Cases.—Before the application abandoned by failure to complete or prosecute can be revived as a pending application it must be shown to the satisfaction of the Commissioner that the delay in the prosecution of the same was unavoidable. If a new application is filed in place of the abandoned or rejected one, a new specification, oath, drawing and fee will be required, but the old model, if any, and if suitable, may be used.

Forfeited Application.—A forfeited application is one on which a patent has been withheld for failure to pay the final fee within the prescribed time. This time, it will be remembered, is six months after the allowance of the application. That is, the final fee of twenty dollars must be paid at some time between the date of allowance and six months from said date. When the patent has been withheld by reason of the non-payment of the final fee, any person, whether inventor or assignee, who has an interest in the invention for which such patent was ordered to issue, may file a renewal of the application for the same invention, but such second application must be made within two years after the allowance of the original application. In such renewal the oath, petition, specification, drawing and model, if any, of the original application, may be used for the second application, but a new fee will be required. The second application will be regarded for all purposes as a continuation of the original one, but must bear date from the time of renewal, and be subject to examination like the original application. Copies of the files in forfeited and abandoned applications may be furnished when ordered by the Commissioner of Patents. The requests for such copies must be presented in the form of a petition, properly verified, as to all matters not appearing of record in the Patent Office.

CHAPTER VI.

INFRINGEMENT—INFRINGING INVENTIONS AND
ACTS—REMEDIES FOR INFRINGEMENT.

The scope of the book does not include in detail such subjects as infringements and interferences, particularly the former, because the subject is so large, the questions arising within it so many, and the nature of the questions requires such discrimination and so many citations that it is impossible in the present work to go into the matter except in a very general way. The main purpose is to tell what to do with inventions and how to make money out of them. But even in treating of this subject, matters relating to the prosecution of patent applications and questions of infringement must be, in a measure considered, to make the work complete and to make such matters fairly familiar to the inventor.

Definition of Infringement.—The grant and issue of a patent gives to the patentee the exclusive right to make, use and vend the patented invention throughout the United States and the territories thereof, during the period for which the patent has been granted. The right is exclusive, and is invaded by one who manufactures, by one who uses only, or by one who sells, presuming, of course, that such person is not authorized by the patentee.

This matter of infringement is apt to be the chief injury against which the inventor can complain, although there are other wrongs against the patentee which cannot be gone into to any extent.

To constitute an infringement, the infringing thing must be the same, or substantially the same, as that covered by the patent. It makes no difference what the patentee may think he has. The patent, as a matter of fact, covers only the patentable subject-matter which is specifically claimed, and which is capable of infringement by an invasion of the inventor's rights in the invention. It follows, then, that there cannot be an infringement, unless the patent clearly discloses the invention—that is, discloses it to such an extent that it may be practiced by one skilled in the art from a study of the specifications, and drawings if any, without regard to any other help except his skill, which need not be above the average.

The invention must be new, and, to a certain extent, useful, to render the patent valid. But supposing the patent to be valid, the next question is, does the alleged infringement come within the scope of the claims of the patent? It is not necessary that it infringe all the claims, but if it infringes any one claim, it is an infringement within the meaning of the law. A patent may have a great many claims, and a majority of them may have no bearing on the infringing article, but if a single one would include in its terms the thing alleged to infringe, then the court will hold that the patentee's rights have been infringed.

Each claim is independent, and must stand or fall of itself. It is read and construed in connection with the descriptive part of the specification, but it is not affected by the other claims, although it may sometimes be inferred that a claim should not have a given meaning or construction because another claim in the same patent has a more clearly expressed meaning that is not consistent with the proposed construction of the claim. It must be borne in mind that a patent does not cover a result, but only the means of producing this result, whether this means be in the nature of a process or a mechanism.

The first question, then, which arises concerning an infringement implies means substantially identical with those of the patent. The two things—that is, the patented invention and the alleged infringement, may be essentially different in many ways, but the alleged infringement has a function or mode of operation or combination of elements which are substantially like those in the patent, so that if the terms of the claim would include the said means, then an infringement exists.

The infringing article may not even have the same elements, but may have well-known equivalents. Equivalent, within the meaning of the patent law, is anything known to exist at the time of the granting of the patent which can be substituted for a given part or parts in the patent and produce the same effect—that is, an element or part that without invention can take the place of another element embodied in the claim of the patent. Well-known examples of equivalents are a spring for a clock movement instead of a weight, a gearing for driving mechanism as a substitute for belts and pulleys or a friction wheel. This similarity of means being established, then, the variations in shape, size, capacity and materials are immaterial.

It has been said that the identity of the alleged infringement with the patented invention is not to be determined by its condition merely at the time of its original construction—that is to say, the infringing thing may not closely resemble the patented invention, but may, when put to use, develop features that perform a function of the patented invention by the same mode of operation. If it performs this result, then it will be held to be an infringement. As heretofore mentioned, the patent grants the exclusive right of making, using and selling the patented invention, these words being intended to cover every method by which the invention can be made valuable by an infringer, and any person who participates in any wrongful appropriation of the invention becomes an infringer.

Not only is it true that the maker, user or seller is an infringer of the patented invention, but if different parties conspire to each make a part of the patented machine or article only, which parts are intended and shall thereafter be combined to produce the patented invention, then each is held to be an infringer—that is to say, supposing a claim covers three pieces of mechanism: A makes one piece, B a second and C a third. These pieces are sold to D, who combines them, making thus a complete article. If the foregoing has been done with the intent to defraud, then all the parties to the conspiracy are infringers. Any person who participates in making, using or selling is guilty of infringement, and liable to the owner of the patent, although an employee who simply works in connection with the making, selling or using without intent to defraud the patentee is not liable; but his employer is. To constitute an infringement there must be some act in derogation of the patent owner's rights, and no matter how strong the presumption, it is not an infringement to possess, expose for sale or advertise the patented invention.   It must be actually made, used or sold to constitute an infringement.

Intent.—The matter of intent is of some importance where a person makes a part only of the invention, because he does not infringe unless he knows that this part is to be combined with some other part to make the infringing act complete, but the actual intent of the real infringer—that is, one who makes, uses or sells the patented invention is of no importance.   He is supposed to have notice by the publication of the patent and by the marking of the invention by the owner of the patent.

Marking Patented Article.—In order that no advantage may be taken of the infringer's ignorance, the law requires the patent owner to mark, if possible, the patented invention, stating that it is patented and when the patent was issued.   Unless the patented invention, if capable of being marked, is so marked, the patent owner can only recover damages arising after actual notice to the infringer, and such notice must be proved.   Although if it can be proved that the infringer really knew that the article was patented, then he will be liable.   It is not practicable to mark an art or process, and therefore the publicity given by the publication and the record of the patent in the Patent Office is supposed to be sufficient notice to the public.

No Infringement Unless Patent Has Issued.—It is a common practice to.market a new invention or an alleged new invention, and mark it "Patent applied for" or "Patent pending," which is a good practice if it is necessary to market the invention before the patent issues, because it may deter others from going to the expense of engaging in the manufacture, use or sale of the article, if they have reason to believe that they will be permitted to do this for a short time only, but one has no remedy for infringement unless the patent actually issues. A patent is a creature of statute, and the patentee has no legal right until the statute gives it to him by the actual grant and issue of a patent, therefore, he cannot set up ownership and proceed against an infringer, until he is in possession of his full rights.

Marketing an Invention Before Issue of Patent.—In this connection it may be well to call attention to the fact that in some foreign countries the patent will issue to the first applicant, even if he is not the true inventor, so that if a person has a valuable article he takes chances in placing it on the market in the

United States before the issue of the patent, and before he has applied for patents in foreign countries, if he contemplates making such applications, for another, being thus placed in possession of the invention, may make application for and secure foreign patents thereon. This subject of foreign patents and their value to the inventor will be gone into fully in another chapter.

Infringement After Expiration.—It is a general rule that the infringing act must be complete during the life of the patent, but there is an exception to this. The patented invention may be made in large quantities secretly during the life of the patent with the intent of marketing it as soon as the patent expires. It has been held that in such case the infringement is actionable after the expiration of the patent, because it would have been impossible to discover it before.

Extends Only to the United States.—As the patent is only issued for the United States and Territories, the infringement must take place within this area in order to be actionable. To make use of or to sell the patented invention within this area, even though it is for sale abroad, is an infringement. And to make, use and sell the invention on an American vessel sailing on any seas is an infringement. But it is not an infringement to make or use the invention on foreign vessels if it is lawfully obtained abroad even though they may be in American waters.

Infringement by Government.—While the patent is granted by the United States Government, still the government or any official thereof has no right to the invention and cannot make, use or sell it without liability, though in cases of public emergency the government may appropriate the invention, but it will have to give the inventor reasonable compensation.

Public Corporations.—A municipal corporation is liable for infringement the same as a natural person and if the infringement is done by officers of the corporation and the corporation reaps the result, it becomes liable therefor. But a municipal corporation is not liable for an infringement committed by a contractor on its public works, nor for the use of patented articles by him, or the use by him of infringing processes.

Private Corporations.—A private corporation is liable for infringement, if the infringing act is done by any of its agents or employees, so long as the infringement is done for the benefit of the corporation and is directly or impliedly ratified by the corporation. To what extent the individual stockholders and officers share, in their private capacities, in the liability for infringing acts is still somewhat of a question and embraces too many fine points to be considered in a work of this character.

Joint Owners of a Patent.—Joint owners of a patent are held to be tenants in common of the patent right and either has the right to make, use and sell the patented invention and he is not liable to his coöwners. A joint owner may, as we have already seen, even go to the extent of disclaiming as to some part of the patent without affecting the rights of the others; he may alienate his interests without the consent of the other joint owners and, in general, may act independently unless the several owners have by some contract specifically defined the relative rights. He can, in the absence of such contract, make any license not exclusive under the patent and, in general, can handle the patented invention as he sees fit and is not liable to an accounting of the profits.

Assignor, Etc.—If one makes an assignment or grant of the entire interest in the patent or the entire interest for a certain territory he becomes a stranger to the patent either in toto or through the specified territory and he is an infringer if he makes, uses or sells the patented invention, just as though he never had any interest in the patent. So a licensee or other person having certain specified rights under the patent either to make or to sell or to use or to do all or any of these

acts in a specified territory, infringes if he makes
any other use of the patent than that which the
license gives him. A person having the license
to sell in a given territory, can sell in that territory, even though he knows that the article
sold will be taken into other territory which is
not included in his license.

Carriers.—A railroad corporation or other
common carrier may be an infringer of a
patent, not only by making, using or selling,
but if the carrier conspires with another to
transport the patented invention out of the
jurisdiction of the United States, where it
may be sold, the carrier is guilty of infringement. Finally, it may be stated generally that
any person, public or private, natural or cor-
porate, who makes, uses or vends a patented
invention is an infringer and as such is liable
to the owner of the patent.

Combination Claims.—It has already been
pointed out that a combination claim covers the
coöperating elements included in an operative
device, or if the claim is for a compound, the
combined ingredients, and that, therefore, only
the combined things are covered by the claim,
while the individual elements are not covered.
Therefore, if one leaves out an element of the
combination or adds an element so as to change
the function of the combination, he is not an
infringer, but if one substitutes for one element

a well-known mechanical equivalent, if the invention comprises a combination of mechanical elements, he is an infringer.

Art or Process.—An art or process is not dependent upon the apparatus by which it is carried into effect or on the results which it produces ; therefore, the claim does not usually cover any mechanism which may be employed in carrying out the art and it does not cover the resulting product of the art. An art usually consists of several steps which are enumerated in the claim, and one to infringe the claim must carry out these several steps, for like leaving out an element of a combination so leaving out a step of the art will avoid the claim. To be an infringer one must practice the whole art, that is, he does not infringe the claim to an art if he only practices some of the steps in the claim, or uses the mechanism described for carrying out the art, or makes use of the result of the art.

Manufacture.—A new or improved article of manufacture is the new or improved thing itself, as distinguished from a machine or means of making a manufacture, as, for example, an improved chair, a shoe, a toy bank. To infringe a patent on such an invention, the article itself or one substantially the same must be made. The manner in which it is made, or the machinery used in its construction, or generally even the material used makes no difference. The actual invention must be made, or used, or sold to constitute an infringing act.

Composition of Matter.—As before remarked, a composition of matter is covered by a claim which sets forth the several ingredients combined to make the composition. Such a claim is really then a combination claim, and such a claim would be infringed if a similar composition is made or if, instead of the ingredients specified, well-known equivalents for them or any of them are substituted. So, ordinarily, leaving out one of the ingredients or adding an element, if the latter really changes the nature of the composition, will avoid the claim.

Design Patents.—Design patents relate to the shape, configuration or ornamentation of the thing and the claim refers to the appearance of the invention rather than to its composition or structure. If a similar invention is made, it infringes the patent on the design, and whether or not the invention is similar is not so much a matter of expert testimony as of ordinary opinion. If the alleged infringement so closely resembles the patented design that an ordinary person would be deceived and would purchase it for the patented design, it is an infringement.

Remedies for Infringement.—To go in detail into the remedies for infringements would necessitate an exhaustive consideration of the nature and title of patents, the jurisdiction of courts, the character of pleadings, the competency of witnesses and a hundred other things which are not within the scope of this work. The law provides ample remedies for infringement and any competent lawyer will know how to proceed according to the facts in each particular case. He will usually bring an action in equity before a competent court and ask for an injunction to restrain the infringer from using the patented invention and that the infringer be compelled to render an accounting. A preliminary injunction is not usually granted unless the patent in issue has been already litigated and sustained. In equity cases, the court, after considering the testimony, decides whether or not there is an infringement, and if there is an infringement, grants an injunction restraining the defendant from further infringements, and orders an accounting, during which proceeding it is ascertained as nearly as possible how much the patent owner has been damaged, and the court can, in its discretion, increase the actual damages and award costs to either party or divide the costs as it sees fit and proper.

CHAPTER VII.

One who has followed the Official Gazette of the Patent Office for several years must have noticed the increase in the number of trade-marks registered. The registration of a trade-mark is to a certain extent in the nature of a patent, that is to say, the government issues to the owner of the trade-mark, a certificate of registration under the seal of the Department of the Interior and signed by the Commissioner of Patents, which is prima facie evidence that the owner has an exclusive right to the use of the mark for a term of thirty years with the privilege of renewal. The right to the trade-mark is practically perpetual and the value of the trade-mark increases with its years. Unlike a patented improvement or invention, the trade-mark is not superseded by improvements, but, as just remarked, its value constantly grows if the goods to which it is applied are of any value. Unlike the inventor, too, the owner of the trade-mark is not obliged to make a race for the Patent Office and put in an early application to protect his rights, and his right outside of the right of registration is not one of statute, but is a common law right. Therefore, a person can bring a suit for infringement of a trade-mark in the courts of any state or territory in the Union.

What is a Trade-Mark.—A trade-mark is some distinguishing mark which a person places on his goods or on the package containing the goods to distinguish them from the goods of others. The trade-mark must be appropriated by the owner or it may be purchased from another, and in such case it goes with the business and good will so far as the particular description of the goods is concerned.

Characteristics of a Trade-Mark.—A person can adopt a trade-mark for any line of goods and it may be in the nature of any distinguishing mark, such as an emblem, the figure of a bird, or even a number, the peculiar arrangement of colored matter on a label or package, the arrangement of certain colored threads in a fabric, if the mark be applied to a fabric, and, in fact, any distinguishing mark whatever, so long as it is unique and arbitrary or fanciful, and original, either with the user or with those from whom he derives his title.

It is a general rule that the trade-mark to be valid must not be descriptive, for instance, a man might use the words "yellow soap" and in such a case the word "yellow" would not be a trade-mark, as it would be either descriptive of the soap or a palpable untruth, so that it would not be valid in either case. It is evident that no one should have an exclusive right to the use of a descriptive adjective. One might use the word "Gold" soap and the word "Gold" would probably be registered, as it would not be descriptive of the soap, but would be suggestive of its good qualities and possibly of its color. One could not appropriate the word "Best" as a trade-mark, because his goods might not be the best and the mark would simply indicate their character or quality. In a very recent case, the Commissioner of Patents held the words "Bromo Soda Mint" to be descriptive and therefore non-registrable, because soda mint is a well-known article and any one would gather from reading the title that soda-mint was combined with bromin or bromid. The courts might not hold such matter to be descriptive, but so far as this question is concerned, the criterion is this : Is the alleged trademark merely descriptive of some quality, color or other characteristic of the goods or is it merely suggestive ? One has a right to select a word for his trade-mark which is suggestive and perhaps this is often the best kind of a mark for particular kinds of goods, but it must not be descriptive.

A very common trade-mark is the word

"Royal" as indicative of goods of the highest character. The word "Queen" is often used for a similar reason and both words have been repeatedly held good trade-marks. Sometimes a person adopts a word which is really descriptive but is used in a foreign form, that is, a foreign word, purely descriptive, is taken, without change into the English and used as a trade-mark, but such a mark has been held as not valid, because a person who understood the meaning of it would see that it was just as descriptive as if used in English. Sometimes words which are descriptive are combined in a single word. Such words do not usually make a valid trade-mark, but they are frequently phonetically and fancifully spelled and in such cases they are usually a valid trade-mark. Descriptive words are used in some instances in connection with a figure or emblem, and in such case the whole may be a valid trade-mark, but it is rather the figure or emblem which lends validity to the mark than the arrangement of the words. One can use an arbitrarily selected word which is a common word, so long as it is not descriptive of the article or is not a geographical word. But one cannot use a word even though it is not descriptive or geographical but is suggestive, if it suggests something which is not true in fact; that is to say, one might use some suggestive term which would indicate that the article on which the mark was used was made of fruit when as a matter of fact no part of fruit composition entered into the article and in such case the trade-mark would be deceptive and hence fraudulent and invalid.

Geographical Name.—One has no exclusive right to use a geographical name, although it may happen that the word may have a popular meaning which is out of all proportion to its geographical meaning. Take the well-known term Trilby, made popular by Du Maurier. It is thought that there are several insignificant towns somewhere in the West which have been given this name. If there are it would be wrong, of course, to deprive one who did not live in the town from the use of the term as a trade-mark, because a vast majority seeing it would associate the word with the novel and never think of the town. But, generally speaking, the rule is that a geographical name is not a valid trade-mark; for example, it would be a great wrong to allow a person to have the sole right to use the word "New York" when there might be thousands of others located in New York making similar lines of goods.

Registration of Trade-Mark.—The proprietor or owner of a trade-mark, whether it be a person, firm or corporation, has a common law right to the use of the mark, which right can be enforced in the ordinary courts, but the object of registration is to give the Federal Courts jurisdiction in trade-mark cases and make an easier remedy for infringement of the mark. This seems rather necessary in view of the fact that there are so many States in the Union so intimately associated in trade matters. A person can register his mark as soon as it is in lawful use, but it is not necessary for him to do so, for if he and those from whom he derived title have used the mark for one hundred years, his right to registration will be so much the better. But in view of the practice and the necessity which may arise to protect his mark it is advisable to have the trade-mark registered in the Patent Office. Not only for this reason, but for the further reason that if the trade-mark is registered and the certificate issued for it, the matter is given more publicity and there will be less liability of infringement by an innocent user of the mark.

Certain Requisites of Trade-Marks.—A trade-mark is unlike a patented article in many ways as already stated and, further, in this, that under the United States law, the owner of a patent is not obliged to put his patented invention in use to maintain his right to the patent, but a trade-mark is good only so long as it is used, and it must be used continuously by the owner in business. He need not make sales of goods bearing his trade-mark every day or every week or every month, but he must have it in constant use and must have for sale the goods bearing the trade-mark. If he neglects for any unreasonable time to use the mark he thereby forfeits the right to the same and another person can adopt it for the same line of goods.

Trade-Mark Valid Only on Certain Goods.—A trade-mark is valid only on the goods on which it is actually used. If a person dealing in groceries uses the word " Star " or the representation of a star on canned goods, but does not use it on other articles, he can then maintain his right to the mark only on canned goods, and another person can use the same on starch, sugar, flour, or other groceries or goods. If a person actually uses the mark on a variety of goods, he is entitled thereby to the exclusive use of the mark on such goods, but on these only. A person may have a trade-mark for cotton goods, but this will not prevent another person from using the identical mark on woolens, another on silks, etc. To entitle a person to registration of his trade-mark, there is a foolish provision of our law, which will probably soon be changed, or at least should be, which requires the proprietor or owner of the mark to make oath that he uses the trade-mark in foreign commerce or in trade with an Indian tribe, before he is entitled to registration. As our commerce is chiefly interstate, the absurdity of this requirement is manifest. Persons owning a valid trade-mark and wishing to register it, therefore, must, before doing so, use their mark to a certain extent in one of the ways above stated, that is, he must sell goods bearing the mark to an Indian tribe or ship them commercially to Canada, England, or some other foreign country.

Interfering Trade-Marks.—The rule as to interferences in trade-mark applications, so far as the question of registration is concerned, is practically the same as to interferences between patent applications, but the question as to whether or not one has been diligent in registering does not usually enter into the case. The person who is the first bona fide and continuous user is the one entitled to registration. For example, suppose that a person applies for registration for a trade-mark which would be registered were it not for the fact that some other person had appropriated the mark for the same line of goods and had already registered the mark in the Patent Office. This registration may have taken place years before, but if the second applicant makes an affidavit that his trade-mark has been in constant use since a time antedating the alleged use by the registrant, he can be placed in interference with the registrant, and then proceedings will be instituted to determine who was really the first and continuous user and to him will be given the decision. If the second applicant proves to be the first and lawful user, then registration will be granted him, notwithstanding the fact that registration has already been granted to another.

Value of Trade-Marks.—In European countries there are trade-marks existing which have been in use for generations and which have come to be of immense value and this is getting to be the case in America. As a country grows older and commerce more extensive, the value of some distinguishing mark on some certain line of goods becomes greater and greater. There are already hundreds of trade-marks used in the United States which are of almost fabulous value, and it is not unusual for a trade-mark to be worth more than a valuable patent. It behooves a person, then, who is manufacturing good goods and proposes to continue such manufacture, to adopt some good mark, if he has not already done so, and have this mark registered in the Patent Office. After a time, people in the trade will then call for such a brand of goods and this brand eventually becomes extremely valuable and identified with the manufacturer. There is a chance for great discrimination in the adoption of a trade-mark and one who gets a happy idea and adopts a valid and unique mark is at once in possession of something which will materially assist in making his business a success.

What is an Infringement of a Trade-Mark.—If the trade-mark is a word, it is infringed by one, other than the owner, who uses a word so similar that an ordinary person would be deceived and so buy the infringing article in mistake for the real one. It has been held also that if the word does not in appearance resemble the trade-mark very closely, but has a similar sound when spoken, then it is also an infringement. If the trade-mark is something other than a word, it is infringed by anything which so closely resembles it as to deceive the ordinary purchaser and lead him to mistake one for the other. The purchaser is not expected to use more than ordinary care or to make any special examination, but if under ordinary circumstances he would mistake one article for the other, then one is an infringement of the other mark.

Remedies for Infringement.—The remedies for infringing a trade-mark are practically like those for infringement of a patent right, except that the trade-mark owner can apply to either the Federal or State courts, according to whether or not the trade-mark has been registered. If he proves infringement he can get an injunction restraining the infringer from the use of the trade-mark and can also recover damages according to the circumstances of the case.

Prints and Labels.—The word print, as used in the act providing for registration, is defined as : "An artistic representation or intellectual production, not borne by an article of manufacture or vendible commodity, but in some fashion pertaining thereto, such, for instance, as an advertisement thereof."

A print is registrable both in the Patent Office and with the Librarian of Congress, according to whether it belongs to an article of manufacture in the one case or pictorial illustrations or works connected with the fine arts in the other. A print to be registered in the Patent Office must relate or belong to, but not be borne by, an article of manufacture or vendible commodity. No prints were registered in the Patent Office prior to 1893, and less than one hundred had been registered up to and including 1898.

The word "label," as referred to in the act relating to prints and labels, is defined as : "An artistic representation or intellectual production impressed or stamped directly on the article of manufacture, or upon a slip or piece of paper or other material to be attached in any manner to the manufactured articles, or to bottles, boxes and packages containing them, to indicate the contents of the package, the name of the manufacturer or the place of manufacture, the quality of goods, directions for use, etc." By articles of manufacture to which labels are applicable is meant all vendible commodities produced by hand, by machinery or by art. No label can be registered in the Patent Office unless it properly belongs to and is to be borne by an article of commerce, and though registration of labels was refused for several years, they began to be allowed again in 1896.

To entitle the proprietor of any print or label to registration, the applicant must sign the application and there must also be filed in the Patent Office five copies of the print or label, one of which, when the print or label is registered, is certified under the seal of the Commissioner of Patents and returned to the proprietor. The certificate, like a copyright, continues in force for twenty-eight years and like a copyright can be extended for a further term of fourteen years if the second application is filed within six months before the expiration of the original term and the other regulations with regard to original applications are complied with. Within two months from the date of said renewal, the applicant must cause a copy of the record thereof to be published for four weeks in one or more newspapers printed in the United States. The fee for registering a print or label is six dollars. If the Examiner at the Patent Office refuses registration, the applicant can petition the Commissioner without fee and have the examiner's decision reviewed. Like a trade-mark, the print or label is infringed by the un-lawful use of the print or label, or a very similar one, by another than the owner thereof.

The act providing for the registration of prints and labels is construed to entitle them to registration without an examination as to their novelty, but though registration may not be re-fused because of a similar print or label, the practice of the Patent Office is to require all trade-mark matter in the label to be first regis-tered as trade-marks.

CHAPTER VIII.

COPYRIGHTS.

Many people have a notion that the Copyright Law is intended for the benefit of inventors, tradesmen and manufacturers, and that a label, a print, a trade-mark, or even, sometimes, a process or mechanism can be protected by copyright. This is a mistaken idea. The Copyright Act is for the protection of purely literary or artistic productions, as a book, a musical composition, a picture or a statue. Perhaps the law, as published by the Librarian of Congress, is, in this regard, the best guide, and is substantially as follows:

SECTION 4952.—(. . . . The) author, inventor, designer or proprietor of any book, map, chart, dramatic or musical composition, engraving, cut, print, or photograph or negative thereof, or of a painting, drawing, chromo, statue, statuary and of models or designs intended to be perfected as works of the fine arts, and the executors, administrators or assigns of any such person shall, upon complying with the provisions of this chapter, have the sole liberty of printing, reprinting, publishing, completing, copying, executing, finishing and vending the same ; and, in the case of a dramatic composition, of publicly performing or representing it, or causing it to be performed or represented by others. And authors or their assigns shall have exclusive right to dramatize or translate any of their works for which copyright shall have been obtained under the laws of the United States. (. . . .)

In the construction of this act the words "engraving," "cut" and "print" shall be applied only to pictorial illustrations or works connected with the fine arts, and no prints or labels designed to be used for any other articles of manufacture shall be entered under the copyright law, but may be registered in the Patent Office. And the Commissioner of Patents is hereby charged with the supervision and control of the entry or registry of such prints or labels, in conformity with the regulations provided by law as to copyright of prints, except that there shall be paid for recording the title of any print or label, not a trade-mark, six dollars, which shall cover the expense of furnishing a copy of the record, under the seal of the Commissioner of Patents, to the party entering the same. (. . . .)

Copyrights are granted for the term of twenty-eight years from the time of recording the title thereof, and in the manner hereinafter directed, and within six months before the expiration of the first term, the author, inventor or designer, if he be still living, can renew the copyright for the further term of fourteen years, or, if he be dead, his widow or children shall have the exclusive right to the copyright, and can get a continuance thereof. Copyrights carry the exclusive right to the property covered, and are assignable in law and by a suitable instrument in writing, and the assignment must be recorded in the office of the Librarian of Congress within sixty days after its execution. If it is not so recorded, it will be void as against any subsequent purchaser or mortgagee, for a valuable consideration, who has no notice of the previous assignment. Before the copyright is complete, it is necessary to make a deposit with the Librarian of Congress of the title or description, also to deposit two copies of the complete thing in its best form when first published, and if the owner of the copyright is a foreign resident outside of the United States, the production, if printed, must be from type set within the United States. The copyrighted matter cannot be imported, except in a few instances, which will be hereinafter given. As to these details, Section 4956 of the Revised Statutes is very specific, and is as follows:

SECTION 4956.—"No person shall be entitled

to a copyright unless he shall, on or before the
day of publication, in this or any foreign country, deliver at the office of the Librarian of
Congress, or deposit in the mail within the
United States, addressed to the Librarian of
Congress, at Washington, District of Columbia,
a printed copy of the title of the book, map,
chart, dramatic or musical composition, engraving, cut, print, photograph or chromo, or
a description of the painting, drawing, statue,
statuary, or a model or design, for a work of
the fine arts, for which he desires a copyright;
nor unless he shall also, not later than the day
of the publication thereof, in this or any foreign
country, deliver at the office of the Librarian of
Congress, at Washington, District of Columbia,
or deposit in the mail, within the United States,
addressed to the Librarian of Congress, at
Washington, District of Columbia, two copies
of such copyright book, map, chart, dramatic
or musical composition, engraving, chromo,
cut, print or photograph, or in case of a painting, drawing, statue, statuary, model or design
for a work of the fine arts, a photograph of the
same, provided that in the case of a book,
photograph, chromo or lithograph, the two
copies of the same required to be delivered or
deposited as above, shall be printed from type
set within the limits of the United States, or
from plates made therefrom, or from negatives,

or drawings on stone made within the limits of the United States, or from transfers made therefrom. During the existence of such copyright the importation into the United States of any book, chromo, lithograph or photograph, so copyrighted, or any edition or editions thereof, or any plates of the same not made from type set, negatives, or drawings on stone made within the limits of the United States, shall be, and is hereby prohibited, except in the cases specified in paragraphs 512 to 516, inclusive, in section two of the act entitled, "An act to reduce the revenue and equalize the duties on imports and for other purposes," approved October 1st, 1890; (. . . .) and except in the case of persons purchasing for use, and not for sale, who import, subject to the duty thereon, not more than two copies of such book at any one time; and except in the case of newspapers and magazines, not containing in whole or in part matter copyrighted under the provisions of this act, unauthorized by the author, which are hereby exempted from prohibition of importation, provided, nevertheless, that in the case of books in foreign languages, of which only translations in English are copyrighted, the prohibition of importation shall apply only to the translation of the same, and the importation of the books in the original language shall be permitted." (. . . .) The exceptions above referred to, being included in paragraphs 512 to 516, are these :

512. Books, engravings, photographs, bound or unbound etchings, maps and charts, which shall have been printed and bound or manufactured more than twenty years at the date of importation.

513. Books and pamphlets printed exclusively in languages other than English ; also books and music, in raised print, used exclusively by the blind.

514. Books, engravings, photographs, etchings, bound or unbound, maps and charts imported by authority or for the use of the United States or for the use of the Library of Congress.

515. Books, maps, lithographic prints and charts, especially imported, not more than two copies in any one invoice, in good faith, for the use of any society incorporated or established for educational, philosophical, literary or religious purposes, or for encouragement of the fine arts, or for the use or by order of any college, academy, school or seminary of learning in the United States, subject to such regulations as the Secretary of the Treasury shall prescribe.

516. Books, or libraries, or parts of libraries, and other household effects of persons or families from foreign countries, if actually used abroad by them not less than one year, and not

intended for any other person or persons, nor for sale. (51st Congress, 1st Session, chap. 1244; 26 Statutes at Large, p. 604.)

SECTION 4957.—The Librarian of Congress shall record the name of such copyright book, or other article, forthwith in a book to be kept for that purpose, in the words following: "Library of Congress, to wit: Be it remembered that on the —— day of —— A. B., of ——, hath deposited in this office the title of a book (map, chart, or otherwise, as the case may be, or description of the article), the title or description of which is in the following words, to wit: (here insert the title or description) the right whereof he claims as author (originator or proprietor, as the case may be), in conformity with the laws of the United States respecting copyrights. C. D., Librarian of Congress." And he shall give a copy of the title or description, under the seal of the Librarian of Congress, to the proprietor, whenever he shall require it.

SECTION 4958.—The Librarian of Congress shall receive from the persons to whom the services designated are rendered, the following fees:

1. For recording the title or description of any copyright book or other article, fifty cents.

2. For every copy under seal of such record actually given to the person claiming the copyright, or his assigns, fifty cents.

3. For recording and certifying any instrument of writing for the assignment of a copyright, one dollar.

4. For every copy of an assignment, one dollar.

All fees so received shall be paid into the treasury of the United States : Provided, That the charge for recording the title or description of any article entered for copyright, the production of a person not a citizen or resident of the United States, shall be one dollar, to be paid as above into the treasury of the United States, to defray the expenses of lists of copyrighted articles as hereinafter provided for.

And it is hereby made the duty of the Librarian of Congress to furnish to the Secretary of the Treasury copies of the entries of titles of all books and other articles wherein the copyright has been completed by the deposit of two copies of such book printed from type set within the limits of the United States, in accordance with the provisions of this act and by the deposit of two copies of such other article made or produced in the United States; and the Secretary of the Treasury is hereby directed to prepare and print, at intervals of not more than a week, catalogues of such title-entries for distribution to the collectors of customs of the United States and to the postmasters of all post offices receiving foreign mails, and such weekly lists, as they are issued, shall be furnished to all parties desiring them, at a sum not exceeding five dollars per annum ; and the Secretary and the Postmaster-General are hereby empowered and required to make and enforce such rules and regulations as shall prevent the importation into the United States, except upon the conditions above specified, of all articles prohibited by this act.

SECTION 4959.—The proprietor of every copyright book or other article shall deliver at the office of the Librarian of Congress, or deposit in the mail, addressed to the Librarian of Congress, at Washington, District of Columbia, (. . . .) a copy of every subsequent edition wherein any substantial changes shall be made: Provided, however, That the alterations, revisions and additions made to books by foreign authors, heretofore published, of which new editions shall appear subsequently to the taking effect of this act, shall be held and deemed capable of being copyrighted as above provided for in this act, unless they form a part of the series in course of publication at the time this act shall take effect.

SECTION 4960.—For every failure on the part of the proprietor of any copyright to deliver, or deposit in the mail, either of the published copies, or description, or photograph, required by Sections 4956 and 4959, the proprietor of the copyright shall be liable to a penalty of twenty-five dollars, to be recovered by the Librarian of Congress, in the name of the United States, in an action in the nature of an action of debt, in any district court of the United States within the jurisdiction of which the delinquent may reside or be found.

SECTION 4961.—The postmaster to whom such copyright book, title, or other article is delivered shall, if requested, give a receipt therefor ; and when so delivered, he shall mail it to its destination.

SECTION 4962.—No person shall maintain an action for the infringement of his copyright unless he shall give notice thereof by inserting in the several copies of every edition published, on the title-page, or the page immediately following, if it be a book; or if a map, chart, musical composition, print, cut, engraving, photograph, painting, drawing, chromo, statue, statuary, or model or design intended to be perfected and completed as a work of the fine arts, by inscribing upon some visible portion thereof, or of the substance on which the same shall be mounted, the following words, viz : "Entered according to act of Congress, in the year ——, by A. B., in the office of the Librarian of Congress, at Washington "; or, at his option, the word "Copyright," together with the year the copyright was entered, and the name of the party by whom it was taken out, thus: "Copyright, 18—, by A. B." (. . . .)

That manufacturers of designs for molded decorative articles, tiles, plaques, or articles of pottery or metal subject to copyright may put the copyright mark prescribed by Section 4962 of the Revised Statutes, and acts additional thereto, upon the back or bottom of such articles, or in such other place upon them as it has heretofore been usual for manufacturers of such articles to employ for the placing of manufacturers, merchants, and trade-marks thereon.

SECTION 4963.—Every person who shall insert or impress such notice, or words of the same purport, in or upon any book, map, chart, dramatic or musical composition, print, cut, engraving or photograph, or other article, whether such article be subject to copyright or otherwise, for which he has not obtained a copyright, or shall knowingly issue or sell any article bearing a notice of United States copyright which has not been copyrighted in this country; or shall import any book, photograph, chromo or lithograph, or other article bearing such notice of copyright or words of the same purport, which is not copyrighted in this country, shall be liable to a penalty of one hundred dollars, recoverable one-half for the person who shall sue for such penalty and one-half to the use of the United States; and the importation into the United States of any book, chromo, lithograph or photograph, or other article bearing such notice of copyright, when there is no existing copyright thereon in the United States, is prohibited; and the circuit courts of the United States sitting in equity are hereby authorized to enjoin the issuing, publishing or selling of any article marked or imported in violation of the United States copyright laws, at the suit of any person complaining of such violation: Provided, That this act shall not apply to any importation of or sale of such goods or articles brought into the United States prior to the passage hereof.

SECTION 4964.—Every person who, after the recording of the title of any book and the depositing of two copies of such book as provided by this act, shall, contrary to the provisions of this act, within the term limited, and without the consent of the proprietor of the copyright first obtained in writing, signed in the presence of two or more witnesses, print, publish, dramatize, translate or import, or, knowing the same to be so printed, published, dramatized, translated or imported, shall sell or expose to sale any copy of such book, shall forfeit every copy thereof to such proprietor, and shall also forfeit and pay such damages as may be recovered in a civil action by such proprietor in any court of competent jurisdiction.

SECTION 4965.—If any person, after the recording of the title of any map, chart, dramatic or musical composition, print, cut, engraving, or photograph, or chromo, or of the description of any painting, drawing, statue, statuary or model or design intended to be perfected and executed as a work of the fine arts, as provided by this act, shall, within the term limited, contrary to the provisions of this act, and without the consent of the proprietor of the copyright first obtained in writing, signed in presence of two or more witnesses, engrave, etch, work, copy, print, publish, dramatize, translate or import, either in whole or in part, or by varying the main design, with intent to evade the law, or, knowing the same to be so printed, published, dramatized, translated or imported, shall sell or expose to sale any copy of such map or other article, as aforesaid, he shall forfeit to the proprietor all the plates on which the same shall be copied, and every sheet thereof, either copied or printed, and shall further forfeit one dollar for every sheet of the same found in his possession, either printing, printed, copied, published, imported or exposed for sale; and in case of a painting, statue or statuary, he shall forfeit ten dollars for every copy of the same in his possession, or by him sold or exposed for sale: Provided, however, That in case of any such infringement of the copyright of a photograph made from any object not a work of fine arts, the sum to be recovered in any action brought under the provisions of this section shall be not less than one hundred dollars, nor more than five thousand dollars, and, Provided, further, That in case of any such infringement of the copyright of a painting, drawing, statue, engraving, etching, print, or model or design for a work of the fine arts, or of a photograph of a work of the fine arts, the sum to be recovered in any action brought through the provisions of this section shall be not less than two hundred and fifty dollars, and not more than ten thousand dollars. One-half of all the foregoing penalties shall go to the proprietors of the copyright and the other half to the use of the United States.

SECTION 4966.—Any person publicly performing or representing any dramatic or musical composition for which a copyright has been obtained, without the consent of the proprietor of said dramatic or musical composition, or his heirs or assigns, shall be liable for damages therefor, such damages in all cases to be assessed at such sum, not less than one hundred dollars for the first and fifty dollars for every subsequent performance, as to the court shall appear to be just. If the unlawful performance and representation be willful and for profit, such person or persons shall be guilty of a misdemeanor, and upon conviction be imprisoned for a period not exceeding one year. Any injunction that may be granted upon hearing after notice to the defendant by any circuit court of the United States, or by a judge thereof, restraining and enjoining the performance or representation of any such dramatic or musical composition, may be served on the parties against whom such injunction may be granted anywhere in the United States, and shall be operative and may be enforced by proceedings to punish for contempt or otherwise by any other circuit court or judge in the United States; but the defendants in said action, or any or either of them, may make a motion in any other circuit in which he or they may be engaged in performing or representing said dramatic or musical composition to dissolve or set aside the said injunction upon such reasonable notice to the plaintiff as the circuit court or the judge before whom said motion shall be made shall deem proper; service of said motion to be made on the plaintiff in person or on his attorneys in the action. The circuit courts or judges thereof shall have jurisdiction to enforce said injunction and to hear and determine a motion to dissolve the same, as herein provided, as fully as if the action were pending or brought in the circuit in which said motion is made.

The clerk of the court, or judge granting the injunction, shall, when required so to do by the court hearing the application to dissolve or enforce said injunction, transmit without delay to said court a certified copy of all the papers on which the said injunction was granted that are on file in his office.

SECTION 4967.—Every person who shall print or publish any manuscript whatever, without the consent of the author or proprietor first obtained (. . . .) shall be liable to the author or proprietor for all damages occasioned by such injury.

SECTION 4968.—No action shall be maintained in any case of forfeiture or penalty under the copyright laws, unless the same is commenced within two years after the cause of action has arisen.

SECTION 4969.—In all actions arising under the laws respecting copyrights the defendant may plead the general issue, and give the special matter in evidence.

SECTION 4970.—The circuit courts, and district courts having the jurisdiction of circuit courts, shall have power, upon bill in equity, filed by any party aggrieved, to grant injunctions to prevent the violation of any right secured by the laws respecting copyrights, according to the course and principles of courts of equity, on such terms as the court may deem reasonable.

The act approved March 3, 1891 (51st Congress, 1st Session, chap. 565; 26 Statutes at Large, pp. 1106 1110), in addition to the amendments, noted above, of Sections 4952, 4954, 4956, 4958, 4959, 4963, 4964, 4965 and 4967, provides further as follows:

"That for the purpose of this act each volume of a book in two or more volumes, when such volumes are published separately, and the first one shall not have been issued before this act shall take effect, and each number of a periodical shall be considered an independent publication, subject to the form of copyrighting as above." (Sec. 11.)

"This act shall go into effect on the first day of July, Anno Domini eighteen hundred and ninety-one." (Sec. 12.)

"That this act shall only apply to a citizen or subject of a foreign state or nation when such foreign state or nation permits to citizens of the United States of America the benefit of copyright on substantially the same basis as its own citizens; or when such foreign state or nation is a party to an international agreement which provides for reciprocity in the granting of copyright, by the terms of which agreement the United States of America may, at its pleasure, become a party to such agreement. The existence of either of the conditions aforesaid shall be determined by the President of the

United States, by proclamation made from time to time as the purposes of this act may require." (Sec. 13.)

The following is a list of foreign countries with which the United States have established copyright relations:

July 1, 1891—Belgium, France, Great Britain and her possessions, and Switzerland. (Statutes at Large, vol. 27, pp. 981, 982.)

April 15, 1892—Germany. (Statutes at Large, vol. 27, pp. 1021, 1022.)

October 31, 1892—Italy. (Statutes at Large, vol. 27, p. 1043.)

May 8, 1893—Denmark. (Statutes at Large, vol. 28, p. 1219.)

July 20, 1893—Portugal. (Statutes at Large, vol. 28, p. 1222.)

July 10, 1895—Spain. (Statutes at Large, vol. 29, p. 871.)

February 27, 1896—Mexico. (Statutes at Large, vol. 29, p. 877.)

May 25, 1896—Chile. (Statutes at Large, vol. 29, p. 880.)

The courts of the United States have jealously and strictly guarded the interests of copyright owners and are quick to provide a remedy in case of infringement of the copyright. The whole matter of publication, for instance, is covered by the copyrights except that the title is not covered by a copyright and may be a trade-mark, but the body of the publication is copyrighted and one who, without permission of the owner, publishes any extended part of the matter, either in the identical shape or in a colorable variation, is liable as an infringer. The intent has something to do with this, however, as a person may make use liberally of quotations if he does so and gives credit to the author, that is to say, he cannot make selections and publish them merely as selections from the copyrighted work, but he may write a review of a book, if it be a book, and make liberal quotations so as to make the review readable and intelligible. But he cannot make any commercial use of the copyrighted matter without infringement, and such use is not necessarily in publishing. If the copyrighted matter be a musical or dramatic composition, the owner of the copyright can prevent the public performance of the composition either by speaking or singing, and even if the composition were memorized, he can prevent it from being spoken and has an action for infringement of a copyright as well as a right to relief by injunction. In this connection attention is called to Section 4966 of the Statutes, which is herein referred to and which refers especially to the questions of damages and injunctions.

CHAPTER IX.

TITLE, ASSIGNMENT, GRANTS, MORTGAGES AND LICENSES.

A patent is as much property as a piece of real estate or a chattel. It follows then that, like other property, it may be sold and transferred, and this right is expressly provided for by Section 4988 of the Revised Statutes. There are four well-known transfers of an interest in or concerning patents, to wit: An assignment, a grant, a mortgage and a license.

Assignment.—An assignment transfers the whole or an undivided interest in the patent for every portion of the United States. The assignment must be written or printed and duly signed. No especial form is required so long as the assignment is absolute and the intent clear. It is better, however, to follow, as nearly as circumstances will permit, the usual form, as this has become well-known and its provisions and restrictions thoroughly understood. It is not necessary that the assignment be witnessed, sealed or acknowledged, but it is better to have it witnessed, sealed and acknowledged, as then all the requirements which may contingently arise are met, and it is especially better to follow such forms, particularly as far as acknowledgment is concerned, because it renders the instrument and its execution easy of proof when necessary. This applies to other conveyances of a patent as well as to assignments.

According to the Patent Office practice : "An assignment, grant or conveyance of a patent will be void as against any subsequent purchaser or mortgagee, for a valuable consideration, without notice, unless recorded in the Patent Office within three months from the date thereof. If any such assignment, grant or conveyance of any patent shall be acknowledged before any notary public of the several States or Territories or the District of Columbia or any Commissioner of the United States Circuit Court or before any secretary of legation or consular officer authorized to administer oaths or perform notarial acts, under Section 1750 of the Revised Statutes, the certificate of such acknowledgment, under the hand and official seal of such notary or other officer, shall be prima facie evidence of the execution of such assignment, grant or conveyance." No instrument will be recorded, unless in the judgment of the Commissioner it amounts to an assignment, grant, mortgage, lien, incumbrance or license, or which does not affect the title of the patent or invention to which it relates.

An assignment or other conveyance should identify the patent by date and number as well as by title, or if the invention is not patented but pending in the Patent Office the name of the inventor, the date of the application, the title of the invention, and, if possible, the serial number should be stated. Instruments are sometimes recorded which do not amount to an assignment, grant, mortgage, lien, incumbrance or license, and while such matter should not appear on the records, still if it is there it may be well to take notice of it.

Where assignments are made conditional on the performance of certain stipulations, the Patent Office can have no notice of whether or not the conditions are fulfilled, and so the records will show an absolute transfer, unless the transfer is canceled on the record by the consent of the parties to the instrument or by the decree of a competent court. One can assign an invention and agree to assign all future improvements and the instrument will not be an assignment, so far as the future improvements are concerned, but the contract will be valid, and if the party makes such improvements he can be compelled by a competent court to make an assignment as stipulated in the original contract. It is customary to make assignments of inventions and of the patent which may issue therefor before the patent has actually issued, but, in such case, the assignment must be recorded in the Patent Office at a date not later than that on which the final fee is paid. If the patent is to issue to the assignee, the assignment must authorize and request the Commissioner of Patents to so issue the patent. This may be important to the parties interested for this reason: that if there is any shadow of equities between the parties and it should become desirable to assign the patent, the proposed assignees may raise quibbles concerning the title and hold that the equitable title is in one person and the legal title in the other, whereas if the patent issues to the assignee no such questions can be raised. For this reason it is usually desirable to have the legal and equitable title merged in one person. If a person dies owning a patent, his executor or administrator can make a transfer thereof. If a person makes a transfer of all his property of every kind and description whatsoever, it would include and carry with it his patent rights. The form of an assignment varies with almost every case. There are many nice questions which come up concerning transfers and the safe way for the parties to a transfer is to have some competent patent lawyer prepare the necessary papers.

A Grant.—A grantee acquires by the grant the exclusive right under the patent to make, use and vend and to grant to others the right to make, use and vend the thing patented within and throughout a specified part of the United States, excluding the patentee therefrom. The essential difference between a grant and an assignment is that the assignment conveys the whole interest or an undivided part thereof for the whole territory of the United States, while a grant conveys an exclusive sectional interest, that is, an exclusive interest for something less than that for the whole country. The law relating to assignments relates also to grants, and what has been said in this regard in relation to assignments is true as regards grants.

A Mortgage.—A mortgage of a patent is substantially like a mortgage of any other piece of property. It can, in the nature of things, convey no more than an assignment and the title is a defeasible one, that is to say, if the conditions, as the payment of money at a given time, are complied with, then the transfer becomes void.

The mortgage to be good as against third parties must be recorded in the Patent Office within three months from the date of its execution.

A License.—A license carries a less interest and a different one from any of the foregoing conveyances. A license may be oral, if it can be strictly proved, but is usually written or printed and in such case must be duly signed. A license may convey a right to make or use or vend or it may convey a right to do all three within a certain territory ; it may convey any interest other than an assignment or a grant. A license may be revocable or irrevocable. It may be exclusive. It may cover a small part of the United States or the whole territory thereof. It may be for six months or for the whole term of the patent and it may convey all the above rights and still be a license merely, for it may leave the title in another who has a right to sue, or have the patent reissued or have the right to disclaim under it. Sometimes a license is broad enough to give the exclusive right to make, sell and convey the patented thing throughout the whole territory of the United States for the full term of the patent, except that in case of certain contingencies, the license shall cease and all rights revert to the legal owner. The patent can be subdivided to such an extent that almost any conceivable use or right under a patent can be conveyed by it. What distinguishes it from an assignment and grant is that the whole interest has not been conveyed, but that a certain interest reversionary, or otherwise, remains in the owner. Sometimes a license is merely implied, as in the case already referred to, where an employee makes an invention on the time or with the tools or materials or at the expense of his employer. In such a case the employer has an implied license, which the court will enforce, to use the patented thing. If a license is to be forfeited on any specified conditions, the conditions when they arise will work a forfeiture, but if the license is to be forfeited by the breach of conditions or by certain acts of the licensee, the fact must be clearly set forth in the license itself, or else a decree of a court is necessary to declare a forfeiture. A license, under a patent, is a nice form of contract and must be prepared with great care by some person skilled in such matters.

Warranty.—As to warranty, the rule is practically the same as it is in regard to the transfer of other property. If the assignor expressly warrants his title, he is, of course, liable under the covenant for a breach of warranty. If he conveys all right in the patented invention it has been held that this amounts to a warranty of title, but if he conveys only such rights as he has, he does not become liable, even though it should prove he had no rights, for this does not amount to a warranty of the title. If the assignor undertakes to assign a certain specified interest, then he is held to warrant that he has such an interest to assign.

Undivided Interest in Patents.—It is a

common saying that parties owning undivided interests in a patent are like tenants in common. Either can sell his interest without the consent of the other. Either can grant a license, though, of course, he cannot grant an exclusive license as this would interfere with the rights of the other. But, generally speaking, each can do with his interest what he sees fit. It has even been held that one cannot be called to an accounting by the other, because, as one judge has said: "None of the parties interested has any right to control the action of the other parties or to exercise any supervision over them. It is difficult to see how an equitable right of contribution can exist among any of them unless it includes all the parties and extends through the whole term of the patent right. And if there be a claim for contribution of profits, there should also be a correlative claim for losses, and an obligation on each party to use due diligence in making his interest profitable. It is not, and cannot be contended that these parties are copartners, but the idea of mutual contribution for profits and losses would require even more than copartnership."

CHAPTER X.

Many United States inventors procure foreign patents on their inventions and frequently do so under a misapprehension of the value of the foreign patents and of the conditions, sometimes onerous, which must be complied with to keep such patents in force. Generally speaking, it is not advisable for an inventor in the United States to procure foreign patents unless it be, perhaps, in Canada, because here a manufacturer may enter into competition with parties holding the patent in the United States. Whether or not the inventor shall obtain a foreign patent or patents depends largely on the nature of the invention; largely, too, on his means of exploiting it, and also very largely on whether or not he has or can make connections with people in the foreign countries who may make use of the invention. A United States patent is almost unique in this, that it is granted unconditionally for the term of seventeen years (except in the case of design patents) and the patentee or his assignee is required to pay no taxes or do no work with the invention unless he feels so inclined and in such cases his rights are not jeopardized.

In almost all foreign countries the laws are different from those of the United States. Usually there is a cumulative tax, being nominal at first, and gradually increasing during the life of the patent, while in others there is a fixed yearly tax. Most foreign countries require also that the invention shall be worked in the country within a specified time, and if the taxes are not paid or the invention worked as required then the patent is forfeited. If a person has patented his invention in half a dozen countries, where such conditions prevail, it will be seen at once that it may be a burden for him to meet the requirements. He should, therefore, before making application for patents know whether or not the invention is likely to be in demand in the countries in which patents are to be obtained, and he should also find out whether he will be able to properly place the patents and bring the invention to the attention of the right parties.

Usually the inventor or patentee has more than he can properly attend to in exploiting his invention in the United States. The above are general rules and not always true. The writers have in mind one man who did not make a great success of his invention in America, but who in the last two or three years has made something

like $400,000 out of his patent rights in Great
Britain and a few African countries. Many instances are known where foreign patents have
proved very valuable, but the parties should
have a fairly clear and definite idea of how they
will work the patents and how dispose of them.

Of course, it is known that in Great Britain,
Germany, France, Belgium, and, perhaps in
Austria and Hungary, certain manufactures are
very largely carried on, and if an invention has
had its value proved in America, and if the
owners are prepared to bring it properly to the
attention of the parties in the countries referred
to, it pays him to obtain patents in such countries and he may realize much money from them.

Much depends on the character of the invention. An invention that would be very profitable in France, or Belgium, or Great Britain
might be of little or no value in South America
or Australia. While, on the other hand, there
are certain mining appliances, some kinds of
agricultural instruments, inventions pertaining
to the handling of live stock or natural products, which may be of more value in South
America or Australia than in the thickly pop-
ulated countries of Europe. Certain woodwork-
ing machinery may be profitable in Canada,
Norway, Sweden and Russia, while it would
be absurd to patent the invention in countries
that are not great manufacturing centres and

where wood is scarce. An invention may pertain to the handling of fruit in some way and be valuable for Spain and Italy and France, while it would be without value in northern Europe and so on through the whole list. One must be governed by the nature of the country and whether or not his invention is adapted to meet the requirements of such a country. No general statement as to the cost, taxes, etc., of foreign patents can be given, because these rules vary with almost every country, but they are at the command of any well-informed patent lawyer.

If the new invention relates to a line which has become established and of proved value in the United States, then one can almost certainly interest foreigners in the invention for other countries, but if nothing has been done here and the invention is still in somewhat of an experimental stage and the owner has no special connections abroad, he had better confine his attention to the United States, where there is a wide field and a chance to reap a good harvest if the invention should prove of value.

BOOK III.

CHAPTER I.

WHAT TO INVENT AND HOW TO INVENT.

Under this title we do not propose to be so specific as to tell a man just how to train himself so as to bring forth a good invention or to specifically point out the things which he should invent in order to make money out of his inventions, but it is thought an inventor can be given such advice as will enable him to invent intelligently and prevent him from wasting his energies. There is no better way open to a poor man to acquire wealth, and at the same time confer a lasting benefit upon humanity, than to bring forth and perfect a good invention.

Most inventors invent because they cannot help it. Their minds are so constituted that new ideas are constantly presenting themselves, and they always see chances for improvement in some line or other. There was a time when the typical inventor had a wild eye, long hair and a haggard look, but that day has passed, and the successful inventor of to-day is a keen business man in a way, although not usually capable of looking after the details of a business, but sometimes he is. Some of the wealthiest men in America are men who have begun life poor and who have brought out some important inventions in certain lines of manufacture, have perfected the inventions and have placed them on the market. After doing this, they have kept control of that particular line of manufacture—that is to say, they have not stopped after inventing a good thing and said "nothing more can be done," but have gone on improving and patenting and even purchasing patents of others so as to acquire and absorb the best means to be had in their particular line.

Once an invention has proved to be valuable, then the cost of a few patents, more or less, is immaterial, and it is best to keep that line of manufacture covered by patents to as great an extent as possible. A man may have a broad patent on a machine, and afterward many people may patent improvements on that machine. These improvements may be infringements, and the broad patent would prevent their use without the consent of the first patentee during the life of his patent. But, on the other hand, the first patentee cannot use the infringements without the consent of the later patentee.

There was a time when it was difficult to

classify inventions and find out just what had been done, but now while the patents issued number over 600,000, still it is an easy matter to find the state of the art in any particular line, as all this vast volume of patents is classified and subclassified to such an extent that any subject within the range of patents can be readily searched. There is no occasion, then, for a man to go about his inventions in a haphazard way, and, perhaps, waste years of valuable time as well as much money.

Before one has gone far with an invention, it is, as a rule, advisable to examine the state of the art to find out what others have done in this particular line. There is hardly a public library of any size in America that does not contain partial drawings and claims of existing patents, at least those issued since 1872, and many such libraries contain the full specifications and claims of all patents. One who has the time can, therefore, himself discover the state of the art by going over the matter in the library and comparing his invention or his idea as it exists in his mind or on paper with what has been done before.

If he finds the field is completely covered he can abandon it, but more often he will find that while there may be inventions substantially like his, yet he will get ideas which will enable him, if he is a bright inventor, to carry the art forward further than he originally intended. Comparatively few inventors have the time to do this, and if so, let such a one put his invention in as good shape as he can and send it with a description to some patent lawyer whom he knows, or who is recommended to him, together with a small fee, usually about five dollars, and the attorney will have the art searched and will send him copies of patents, showing devices as near as may be to his, and will further advise him as to the probabilities of getting a patent. It is well to make this search, either by attorney or personally, before going to great length with the invention, because so many inventions have been patented that the inventor may find practically his own ideas already covered, though they may be original with him, so far as he is concerned. Often, and, perhaps, usually the inventor will go ahead, without any attempt to see what has already been accomplished in his line, and, of course, after he has gone to the expense of spending his time and money, and to the further expense of making a patent application, it is a sore disappointment to find that his claims are substantially met. If the line of invention seems important and the inventor wishes to be thoroughly informed on it, he can often secure the whole subclass of patents to which it relates for a comparatively small expenditure.

Recently the cost of patent copies has been reduced so that one can order a single copy for five cents; a subclass, and get the copies for three cents apiece; a class, for two cents apiece, and all the patents issued for one cent each. It will be seen that one can easily know what he has to contend with before he goes very far with his invention. On the other hand, many inventors have already done so much in certain lines that they are thoroughly familiar with the art and know practically just what has been attempted before their invention. In such a case, the proper thing to do is to at once file the patent application. It may be that the invention is of such a nature that one may feel reasonably sure of its novelty, and had rather apply at once than wait for an examination. It may appear also that there is danger of an interference with some other party —that is, that another may file an application for the same thing, and, if there is reason for haste, then it is well to file the application.

What to Invent.—What has been said already in this chapter relates particularly to how a man can acquaint himself fully with the art. That which is most important, after all, is, perhaps, to know what to invent. Some of the most ingenious things have been of no practical value, and there have been some inventions recognized as great inventions which have been of no commercial worth. The average inventor is not seeking fame as much as he is seeking money, and, therefore, he wishes to direct his ingenuity in the best commercial lines. Let him, therefore, when an invention suggests itself to him ask himself first of all, what the demand for it will be if it is successful, as he hopes.

It is not necessary to confine ingenuity to great lines, because some of the greatest commercial affairs of the country are founded on little things, like glove fasteners, matches, toothpicks and woodenware, hairpins, hooks and eyes and a thousand other things, but he should know that whatever his invention is there will be a demand for it if it is up to his expectations so far as structure is concerned. If the invention relates to some staple article of manufacture or consumption, and he can devise machinery or means to cheapen the said article, he is practically sure of a commercial success, because if he can show any manufacturer that he can save him money, he will find a very ready listener and find a class of people ready to meet him if his invention is properly protected. He is certain of being able to put a part of the saving into his own pocket if his invention is properly managed. If, on the other hand, he can make the said article of manufacture so that it will cost no more, but will be really better, then, too, he has something which will pay him well. If not at once, it will in the long run, if the invention is handled as it should be. One of the most successful inventions, coming under our personal knowledge, is of this latter class. The inventor had a means of making a well-known article of manufacture, so that it was a little more desirable than it would otherwise be, and it cost no more than similar articles made without the improvement. The inventor hesitated about making the application for a patent, and even allowed his first application to lapse, but afterward procured his patent and began making his goods. They cost no more than did goods of his competitors, and he soon found that wherever his goods were offered in competition with those of others he received the order, and, as a result, his trade increased by leaps and bounds, and he realized handsome returns from his little invention.

It is a notorious fact that first inventors usually employ complicated means to attain the desired result, and that following improvements usually simplify the means. If one can see a way of simplifying a well-known process of manufacture or a well-known machine, he will usually find, even though he cannot use the improvement independently of the original inventor, its value will be recognized, and he can get a good return for what he puts into the invention in the way of time, ingenuity and money. Inventions to be profitable need not be of either class above referred to, but the inventor should satisfy himself that there is a demand for the invention. Perhaps the invention is a toy. Many such have proved wonderfully remunerative. But he should take some means to find out whether the toy will be a selling one before he goes too far with it. In inventions of this kind it is not always possible to do so, and it is something of a speculation, and a person cannot always tell just how a thing will take until it is tried on the market. Almost every one can call to mind certain toys and games which have been patented, and from which the promoters have made fortunes. The invention may be a design, and be very profitable, but usually this line of inventions is confined to a class of people having more or less to do with the manufacture of artistic articles, though this is not always the case.

Sometimes one will conceive a design for an article of manufacture other than an ornament, or even an ornament which will commend itself at once to those who are engaged in the line to which the design appertains. Frequently the shape given to the invention will be of such novelty as to give it such a new function that the article can be covered by an ordinary seventeen-year patent. But, generally speaking, the inventor wastes his ingenuity if he allows himself to work on articles for which there will be no profitable sale, or which will not in some way affect some line of trade or manufacture sufficiently extensive to give him a good reward. There is no need for him to make such a waste of his energy, because there is plenty of room for the best inventive skill along remunerative lines.

How to Invent.—The inventor, whether of patentable inventions or of those which do not come within the purview of the patent law, is one who is of open mind and is looking constantly for something new and who is never satisfied with what has come to him at second hand. Almost any one can invent, though all cannot be great inventors, and a natural inventor will, ordinarily, do more and better inventive work than one who is not; just as a natural poet will write more and better poems than one who has to labor to bring forth a little rhyme. But any one may see room for improvement.

Do not take things for granted. The steam engine of a generation ago looked a veritable wonder to the people of that time, but the same engine would look crude indeed as compared with one of recent build or with an up-to-date electric motor. Find out the whys and wherefores of things, and see if they cannot be improved. If you see a piece of work being done in a way which seems crude, ask yourself how it could be done in a better way. If a thing does not work to your satisfaction, ask how it may be improved so as to approximately meet your ideas. If a thing is too expensive, study to see how it may be cheapened, and in these ways you may discover something of value to yourself and the rest of the world. Some great inventions have been discovered accidentally, but usually the accident has come during the course of experiments along the line to which the invention relates. If you do not accept conditions of things as being ideal, but look earnestly for improvement, it will surprise you to see how many crudities will come to your attention and how many improvements will suggest themselves.

This being done, then discriminate, and see what suggested improvements are worth following up and what are worth patenting and exploiting. It is not necessary that the invention be in the line of business in which the inventor is engaged. It happens as often as otherwise that a person will see a new machine or a new process for the first time in his life, and will ask why some things are not done in a certain way or will note at once a means to simplify a machine or process, or, perhaps, change it for the better.   All of which goes to show that a man who has his eyes open may see improvements almost anywhere.

CHAPTER II.

INTRODUCTION AND SALE OF PATENTED INVENTIONS.

The average inventor is completely taken up with securing his patent, and after he has received it he finds himself at a loss to know what to do with his invention. Frequently he is a man who has not had much business experience and he does not know how or where to begin, does not know how to sell the invention, to interest capital, or conduct the business generally. The best means of handling the patent depends on the intention of the inventor as to its exploitation. He should know first that he has something worth introducing or else he will be sorry if he tries to do anything with the invention. If he has something that is worthless and succeeds in palming it off on some one, the result will be unsatisfactory in the end. Let him, then, satisfy himself that he has a really good thing and that it is of value. His next step will depend on one of several things: First, Is it his intention to stay in the business to which the patent relates? Second, Is he a prolific inventor? Third, Is his business already established and does the patent simply enhance the value of his business? Fourth, Does he wish to establish a business founded on his patent? Fifth, Does he wish to sell patent rights, that is, territorial rights? Sixth, Does he wish to sell out his invention for cash? Seventh, Is he satisfied to have the invention worked on a royalty?

If the inventor intends to manufacture and control his patent himself and has sufficient capital to work it, then the advice here given would be of no especial value as it would simply be a commercial affair and he would use ordinary business methods to bring his invention before the public. If it is a machine he will manufacture the machine, and use every legitimate means to advertise it and bring it to the attention of the public in the line to which the invention relates. He will, if possible, get the machine at work where it can be compared with others and if it is superior it will eventually make its way. If the inventor has not the capital himself, and this is usually his predicament, he must in some way get some one to put in money with him to promote his invention. In this case let him remember that the money is as essential to develop and work the invention as the invention is to make capital profitable, and he must therefore be willing to give some one a reasonable chance to share in the success of the venture and should not expect too much for his invention, particularly as the invention is usually somewhat in the nature of an experiment which may not prove successful.

Usually the money necessary to properly push the invention should command as great an interest as the invention itself, but the inventor may be at a loss to know how to interest capital even on this basis. The first thing necessary is to properly exhibit the invention. If the inventor takes a somewhat crude drawing or even a good drawing and attempts to explain his invention, he will find the result unsatisfactory. Nine out of ten will say: "Oh, yes, we understand it all right," and perhaps they do, but even if they do, it will be found that the same men will be very much more favorably impressed if the inventor exhibits to them a working model or, better, a full-sized device showing the advantages of the invention.

Men who are working over drawings every day for years are themselves constantly surprised to see how much more favorably they are impressed with the real thing than by any drawing that might be submitted. It is often pretty difficult for an inventor to get money to make a model to properly show his invention, but let him persevere. If he cannot make a model himself, let him find somebody, some friend or other, who will advance the money and take his pay out of the profits or accept a small interest. The invention must be presented properly or he never can secure satisfactory results. If he is satisfied that the invention is valuable, he may make great sacrifices to put it in shape. We have in mind one inventor who made an immense fortune out of one of the best-known steam appliances in the country, and have it directly from his son that he was forced to sell the bed from under him to get the money to make his model and show his invention. The inventor had a terrible experience, but he was successful, and there is nothing like success. We believe with Emerson that the law of compensation is sure to come in somewhere, and the inventor must give some sort of an equivalent for what he receives. If he is compelled to economize and struggle and make many sacrifices it amounts to nothing if his efforts are crowned with success. All great successes usually come as a reward to great efforts. This is recognized as a rule and is especially true in regard to inventions. Do not therefore be discouraged. Do not lie by and expect some one to hunt up the invention and buy it or push it, but persistently bring it to the attention of some one who can help you. The inventor need not feel any modesty about this, for if the invention is good he is doing a favor to the party whom he interests. When he has succeeded in getting his business started, he must then be eternally vigilant and see that the invention is kept up-to-date, that is, he must, if possible, anticipate improvements and cover every novel feature by a patent, as when the business is once started its profits depend upon the monopoly which the patent gives and the cost of a few patents, more or less, under such circumstances is of no consequence. When a business founded on a good patent cannot be started by a few people in their individual capacities, a favorite means is to promote the invention through a joint-stock company or corporation, This subject will be treated in the following chapter.

Is the Inventor Prolific?—Our experience is that real prolific inventors will not, as a rule, be tied down to any one business or line of inventive work. We know many men of this kind who invent almost constantly and will sometimes make a great deal of money out of the sale of an invention, will then go into business and lose it all and will finally bring out some other good invention and make another small fortune. We have in mind many such people as this and do not doubt that many inventors who read this will see that the coat fits them. Such people are very foolish to attempt to manage any business or to be too closely identified with it. To such a one we say: keep on inventing, bring out your invention, put it in good shape to show, then sell your patent rights and keep on inventing and selling. If you are like many prolific inventors, this is the only way you will make any money, and if you are interested in some of your inventions, that is, in a commercial way, let some one else manage the business. One man is seldom good at everything. It is one of the greatest gifts to be able to bring forth good inventions, and a man who can do this successfully and often cannot expect to do everything else equally well, and should be satisfied to reap the reward of his ingenuity by selling the products of his brain without any attempt at commercial exploitation.

Does the Patent Relate to Established Business?—There is a class of shrewd inventors who are successful manufacturers and whose ingenuity comes with the need of making a dollar in the line of business in which they are engaged. These men generally manufacture specialties covered more or less by their own patents, and this is what makes the business profitable. They should, therefore, see to it that they keep the line of manufacture protected and such a man should not think he is the only one who can invent, but should take up any meritorious invention in his line even though it is invented by some one else, because if another invention is nearly as good, it means competition and reduced profits.

Territorial Rights.—As much depends on the way in which a patent is handled as on the invention itself, so far as profits are concerned. Many inventors, after patenting a real good invention, seem to lose interest in the matter and let it drop. There is never a good thing for which there is a demand that cannot be sold. Some quick fortunes have been realized by selling out the rights under a patent, under territorial grants. The inventor is usually capable of giving a good explanation of his invention and showing up its methods. Let him then, after getting his patent, make a model, or if it is practicable, a full-sized device and exhibit it in the different States, counties and towns. Let him engage a few people whom he knows to be honest and active to assist him in this work and he will find that by putting in hard and earnest and persistent work in showing up this invention in different communities, he can in almost every county find a customer who will pay him well for an exclusive county right or he may find a customer who will pay more for a State right. This can be and is done over and over again by enterprising inventors and the result is almost always highly satisfactory. Very often an inventor will make an arrangement

with a manufacturer who will agree to supply
the invention in certain quantities at a certain
price. He may retain an interest in this manufacture. He then sells the territorial rights as
widely as possible and agrees to furnish goods
at a given price. He thus gets a quick profit
from the sale of the patent right and a continuous profit from the manufacture of the goods,
if the sale proves to be considerable. The inventor should not ask too much or at least
should not insist on getting too much for his
territorial rights, but, as a rule, it is better to accept a reasonable offer, even though it seems
low, for the profit is quick and the expenses
light.

Selling a Patent Complete.—If the invention is a good one and the patent reasonably
strong, the inventor can by intelligent and continued effort usually interest some one engaged
in the line of business to which the patent relates. In this case the invention should not be
shown to the party whom it is desired to interest until it can be put in good condition, so
that no explanations or apologies are necessary.
If he can show a good thing and the invention
is capable of speaking for itself, as the saying
is, he will find parties who are interested in it.
If it saves a man money, he is quickly interested. If it bids fair to enlarge his trade, he
is also interested. The main thing is to have it

properly presented. Many inventions which are of little or no value are called to the attention of manufacturers and capitalists, thus making them lose faith in inventions, and sometimes such men, if they really wish to buy an invention, will discredit it; will say boldly that they have found the patent is invalid; will even threaten to make the invention or will state that it does not interest them; when, as a matter of fact, they are really aiming to buy it cheaply. This is the reason an inventor can sometimes act better through a third party, although this depends somewhat on the character of the people with whom he is dealing and whether or not he is in a measure acquainted with them. If it can be brought to the attention of proposed buyers and shown that the invention will probably come into competition with them if they do not control it, they are more likely to buy the invention and will frequently buy it to prevent such competition even if they do not intend to use the invention directly in their business.

There is hardly a large manufacturing corporation in America which does not own scores of patents for inventions which it does not use and some of them own hundreds and even thousands of such patents. If the invention has merit and can be brought properly to the attention of such people it will usually sell, but the inventor should not expect to get the last dollar there is in the invention, and will usually make a mistake if he does not accept a reasonable offer.

The invention can often be sold by judicious advertising. If a well-worded advertisement is placed in a daily paper of a large city, it will usually attract attention, and some of the answers to the advertisement may lead to a sale. Advertisements in obscure papers or papers devoted specially to patents and particular lines are not so apt to be profitable as those placed in well-known daily papers of large circulation. It pays to have a well-written description and a cut of the invention, which can be sent to the parties expressing an interest in it, and the inventor should be able to make a conservative statement as to the probable extent of use of the invention, its sale or saving and the profits likely to be derived from it. In this, as in everything else, a system carried out intelligently and persistently is almost sure to bring good results. There are many ways in which the inventor can sell his invention. But he must try to sell, keep working and follow up the different clues which present themselves. If he does this he is almost certain to sell in the end.

Royalties.—There are many manufacturers, small and large, who are willing and anxious to take up a good invention but who cannot spare the capital to buy the patent outright. Such men will pay a reasonable royalty on something which appeals to them and in the end will make a handsome thing out of the investment, and the result will also be very satisfactory to the inventor. The inventor can usually get a good income by an arrangement of this sort and get very much more out of the invention than if he attempted to sell it outright. Such an arrangement is usually made by giving a license to the manufacturer, which license may be exclusive or otherwise, according to the nature of the case, and the licensee should be willing to pay something in cash as a guaranty that the work will be pushed. He should agree to make and sell at least a certain number per year. He should agree to make returns under oath at stated intervals of the amount of his sales. He should also agree that the records of sale should be open to the licensor. He should further agree that the royalties should be paid at stated intervals and it should be also understood and agreed that if the terms of the license are not complied with, the license is thereby forfeited.

Caution.—It has been told how the aid of a third party may be of service in selling an invention or patent and in a later chapter it has been stated how a promoter can be successfully used in some cases. But the owner of the patent must be on his guard in such matters, for it is the custom, and has been for many years, for agencies and individuals to flood the country with advertising matter which is sent chiefly to patentees and tells how the advertiser can sell patents. As a rule the people sending out this matter are unreliable and seek only to get what money they can from the patentee. It is seldom indeed that most of them effect a sale. It is their custom to require a bonus to cover advertising expenses and other little bills, but this bonus is usually for the sole benefit of the person who claims skill in selling. Where agencies advertise to sell patents, it is usually with the idea of getting the inventors to file through them their patent applications and it is rarely they effect a legitimate sale. There are competent parties who do such work, but their reliability must be ascertained and, as a rule, they do not require any cash payments in advance.

In conclusion, the inventor should not sit down and wait after he receives his patent, but should gird himself for the struggle which has just begun and work hard and constantly until he forces the invention to the attention of some party who will either work it or buy it. This can be done either by his own efforts or by the efforts of some party whom he interests. But, as a rule, the efforts of any third party must be supplemented and augmented by the never-ceasing efforts of the inventor himself. If the invention is good and properly protected by Letters Patent, such efforts as these will eventually effect a sale.

CHAPTER III.

In this chapter it is not proposed to treat of joint-stock companies and corporations *per se*, but to define their peculiar characteristics and powers only so far as they pertain to the subjects of exploiting or selling inventions. If a person has to raise a comparatively large amount of money it is not usually easy to find one or two people who are willing to risk the whole amount or a considerable part thereof in an experimental venture, and it is much easier to find twenty-five people, for instance, who will risk a thousand dollars apiece, than it is to find one man who will risk twenty-five thousand dollars or two who will risk twelve thousand five hundred dollars each. Moreover, when a person in his individual capacity joins another to promote an invention in any way, he will usually become a partner in the enterprise and so become liable beyond the amount of money which he invests. But by becoming a stockholder in a legally organized corporation, he only runs the risk of losing his investment or a part of it and has, perhaps, a better opportunity to get a profit. Usually, too, the inventor or patent owner wishes to make some money out of his patent quickly, as well as to provide himself a continuous income. If a private person puts up money to promote the patent enterprise, he may insist that all the money shall go into the business, whereas by organizing a corporation and selling his invention to it, as described presently, the inventor can usually retain a percentage of the money actually paid in and in any event can retain a large stock interest and his stock can be disposed of in greater or less amounts as the necessities of the case require.

Probably a corporation organized to handle or exploit a patented thing affords the easiest means by which an inventor can market his invention either to get his money out of it or to get it on a business footing. This can be done in several ways. We will suppose that the inventor has a meritorious invention worthy of exploitation and that it requires, say, twenty-five thousand dollars to start a manufacturing company which can properly handle the invention. We will suppose that the inventor can show that with the expenditure of a certain amount and the proper managing of the business, the profits of the concern will pay good dividends on a hundred thousand dollars capital. In some States, the corporation laws require that the whole capital stock of the corporation must be paid in in cash or in a good cash equivalent and the laws are construed to mean that property turned in in lieu of cash shall be actually worth the valuation placed upon it. But there are many others having liberal corporation laws, such as Maine, West Virginia, New Jersey, and several Territories and many other States in which a very small amount can be paid in in cash and the balance in property at such a valuation as the organizers or stockholders fix. Nearly all States and Territories, however, require that the stock at its first subscription and issue must be sold at par. It is not obvious, then, how one can organize a hundred thousand dollar corporation, sell twenty-five thousand dollars' worth of stock at less than par and provide a large stock interest for the patentee or patent owner and still keep within the letter of the law. It is done in this way: We will suppose that an inventor has found parties who are willing to risk some money in the enterprise and will take sufficient stock at, say, seventy cents on the dollar, to realize the twenty-five thousand dollars. He then with several other parties, more or less, according to the requirements of the State in which he organizes, incorporates under a certain name. The incorporators and stockholders at their first meeting adopt a resolution for the purchase of the patent or patents which it is proposed to turn into the company and authorize the proper officers to issue, say, $99,500 worth of the company's stock at par, in payment for said patents, which would leave $500 —to make the full $100,000—which must be paid in cash. But, of course, this amount may be varied and made more or less. The stock being issued to the patent owner and the patent transfer to the company being duly made, the corporation is then in possession of the patent rights with no means of working them and the former patent owner is in possession of $99,500 of stock in a corporation having no money in its treasury and no means of doing business. He then says: "To make my stock valuable and to promote the best interests of the company I propose to give to the treasurer or some other person in trust for the benefit of the company $45,000 of said capital stock, which stock shall be sold to provide a treasury fund for carrying on the company's business." This is done and the stock accepted by resolution, which, like the former one, is spread on the minutes of the company.

The stock at this point has been legally issued at par and the whole capital paid in. This being done, the owners of the stock, whether

the trustee aforesaid, acting under authority,
or the former patent owner, can sell their stock
at any price they see fit, whether it be more or
less than par. The directors of the company
meet and pass a resolution authorizing the sale
of sufficient treasury stock held by the trustee
for the benefit of the treasury, at seventy per
cent. of its par value, to realize the proposed
$25,000. This being done, the stock can be
legally sold to the parties who have agreed to
take it or to others if necessary, and the company is then in possession of $25,000 in its
treasury and a reserve treasury fund of stock
which can be sold if need be. The parties,
agreeing to take the stock at seventy per cent.
of its value, have their stock legally issued, and
the former patent owner is in possession of approximately fifty-five per cent. of the capital
stock. This leaves the inventor in safe control
of the company and if he wishes to sell a few
thousand dollars' worth of stock he can do so
and still retain a safe interest, or, at the proper
time, he can unload entirely, if he prefers to
do so. Unless the party who has been instru-
mental in organizing the company can sell
sufficient stock to satisfy him with the price
obtained for his patent, he should, either individually or in connection with some reliable
friends, retain the controlling interest, that is,
more than fifty per cent. of the company's

stock. If he does not intend to stay with the company, he can sell all the stock possible or wait until the company is well under way and sell it, perhaps, for a greater price, but if the corporation is intended to do a business with which the patentee and associates are identified, he and his friends should retain control.

In almost every corporation, every stockholder is entitled to one vote for each share of stock owned or represented by him. It will be seen then that if the patentee and his associates held less than fifty per cent. of the stock it would be possible for the other stockholders to combine and elect all the directors and other officers of the company which are chosen by the stockholders. It is well understood that practically all the powers of a corporation are left with the directors and if it should happen that the directors and the majority interest referred to should be inimical to the patentee and associates, they might do such acts as would make his interest practically worthless. They could fill all the offices, vote themselves salaries to which they might not be justly entitled, and in many ways work for their own interests to the exclusion of that of the patentee. It is well understood that the controlling interest in a corporation is necessary unless several of the stockholders have some means of knowing that the affairs of the company will be honestly administered in the interest of each and every stockholder. For this very reason, parties are sometimes loath to invest in a corporation unless satisfied as to the control. They may, therefore, hold aloof from investing, if the patentee and promoter of the corporation is to have the controlling interest. In such a case, he can arrange by suitable agreement to have the controlling interest in the stock pooled, and voted in such a way for a definite term that the interests of the incoming shareholders and himself can be equally looked after. Or he can arrange with them that for a certain period each party shall be represented by a certain number of directors in the corporation.

If the incoming or first shareholders insist on having the controlling interest of the company, the inventor may perhaps meet their views and protect himself in this way. He can assign his patent or patents to some person or corporation in trust to hold for his benefit. The trustee can then give to the new corporation an exclusive license to make, use and vend the invention and any improvements thereon, but the license can stipulate that in the event of the corporation being wound up by the Attorney-General, or going into the hands of a receiver, or performing certain specified acts which would be injurious to the patentee, then, in such case, the license shall be forfeited and the patent with

all its rights revert to the trustee. There are
many advantages in handling an invention
through a corporation having a board of able
directors. If the men constituting the board
are good business men, their advice and experience are of great advantage in properly pushing
the invention. Then, as already stated, no one
is individually liable for the debts, defaults or
misdoings of another unless it be a case of
fraud to which he is a party, and no one of
them is liable for the debts or losses of the corporation.

It will be seen that the organization of a corporation is comparatively simple, but the patentee may yet be at a loss to know how to
interest parties to the extent of taking stock.
This must be by his personal efforts, supplemented, as stated in a preceding chapter concerning sales, by the efforts of those who can
help him. Unless he can go to parties known
to him and interest them directly—parties who
will from their knowledge rely on him to a certain extent as well as on the invention—he must
have means to show them that money can be
made out of the enterprise. There is plenty of
capital seeking investment, but it must be made
plain that there is a probability of success. To
show this the inventor should have some means
of properly showing up his invention. If he
has sufficient money to build the machine or

article of manufacture, or to illustrate the process, as the case may be, he should, by all means, do it. It is well then to get up a good statement or prospectus, showing just what the invention is, what its uses are, the probable extent to which it will be used and the profits which should be derived under given conditions. If he can have an illustration of the invention made, it is well to have such illustration on the prospectus. These prospectuses should be sent to parties whom he has reason to think may have money for investment and he should follow this matter up by personally seeing as many persons as possible and explaining to them every feature of the invention. He should also get them, if he can, to make a personal inspection of the working invention, if such has been made. If the inventor is not a man competent to make a good presentation of his case, he had better secure the services of some one who is well skilled in such business, and he can afford to pay him well, but of this matter we shall treat in the following chapter on promoters.

The inventor need not be overcautious. He must bear in mind that if the invention he offers is really a good thing, he is working an advantage to the people whom he interests and he must remember that it takes hard and persistent work to float any enterprise. Let him consider while doing it that if he carries the invention through, he raises himself from, perhaps, a life of poverty to one of comparative comfort. Let him remember the statement attributed to Solomon : "Yet a little sleep, a little slumber, a little folding of the hands to sleep, so shall thy poverty come upon thee as a robber and thy want as an armed man." Of course, it is nice if one can sit down, send out his prospectuses and have investors come trooping up anxious to subscribe to his stock, but things do not happen in this way and as "eternal vigilance is the price of liberty," so eternal hustle is the price of success. Let him go into this matter with enthusiasm and work intelligently and earnestly, even at the risk of boring some few people, and he will be surprised to see what results he can accomplish. One can do anything in reason if he only works hard enough and makes sufficient sacrifices. It is a good scheme to introduce an invention by means of a corporation if the promoter is honestly intent on building up a good business, because, with all due respect to the inventor, he is not usually a man of details, is not always accustomed to the minutæ and intricacies of a commercial business and is much more successful if he has the coöperation of men trained in this line.

In this chapter the term promoter is used to define the one who instigates the organization of the corporation, but the term is used generally and in distinction of the definite promoter who has come to have a recognized place in the commercial world and to whom it is proposed to devote a chapter.

An example has been given of a hundred thousand dollar corporation in which it was proposed to raise $25,000 as a cash working capital. It will be understood from this that the terms and conditions can be varied indefinitely. Perhaps a shrewd inventor who is organizing a corporation has in mind the combining of several other concerns engaged in an analogous business. In such case he should make his corporation relatively large. He may not need more than $25,000 in money ; he may not need even so much as this. But he may have in mind the absorption of perhaps half a dozen concerns already established. The other concerns may have more business and he may have an improved means of carrying on the business so that by combining with them he gets the field and the better means of working it.

We will suppose that he can show that the combined profits of the concern under existing conditions would pay dividends on say two million dollars of capital and perhaps he can show that by the combination which will reduce expenses and by the introduction of his improved means, the concern will pay good dividends on five million dollars. Let him organize his new company for five million dollars, paying in his capital in the manner already set forth, but setting aside a relatively large proportion of the stock for the use of the treasury. This is rather an ambitious attempt and, of course, the man who simply brings in his improved means should not expect to keep the controlling interest in the combination effected by the new corporation. Each party to the combination will, of course, expect a fair representation, but after he has organized, and possibly before, he can arrange with the parties, so that each will turn over its property or at least the management and control of the business to the new corporation, taking in payment therefor a certain amount of the stock, and after this combination is effected the inventor should find himself in possession of sufficient stock in a solid corporation to amply pay him for his inventions and patents which he has contributed. He should also find himself in possession of stock sufficient to well pay him for the labor involved in effecting the combination and, finally, he will have a powerful corporation behind his improvements ready to push them to success.

Preferred Stock.—Many investors who are not speculators would rather feel reasonably

sure of a small income than to have a fair pros-
pect of getting a large income if the latter was
contingent on a chance of failure. Sometimes,
therefore, it is better in organizing a new corporation to provide for a certain amount of preferred stock which shall take the first earnings
for a dividend, even though the common stock
may earn much more than the preferred stock,
or may earn nothing. It will be supposed, as
in the first instance given, that it is necessary
to raise twenty-five thousand dollars in cash.
When the corporation is organized, instead of
providing the treasury fund as suggested, a
portion of the forty-five per cent. of stock devoted to treasury purposes—say twenty-five
thousand dollars of it—is made preferred stock
to draw six or seven per cent. interest, as the
case may be. Then instead of offering the stock
for a relatively low price it is offered at par.
The nature of the stock is this: It is not in the
nature of bonds secured by mortgage on the
property of the corporation, but it is provided
by the articles of incorporation that the earnings of the corporation shall be applied to pay
the fixed dividend called for on the preferred
stock before any such earnings are applied as
dividends on the common stock. There may
be sufficient business in sight to make it evident
that the preferred stock dividends will certainly
be earned. In this case, the stock will readily

sell for par and will be worth much more than the common stock. Preferred stock may sell for par—say one hundred dollars per share—when perhaps the common stock will hardly sell at thirty dollars, but the preferred stock can never earn more than the amount stipulated, say six per cent. If then the corporation is successful and earns in a year enough to pay twenty per cent. in dividends, six per cent. will go to pay the preferred stock dividend and the balance will be dividends on the common stock. If this condition of things occurs then the common stock will, of course, be worth more than twice as much as the preferred stock.

In small corporations it is not customary to issue preferred stock, but there is no special reason why it should not be done and sometimes it affords the readiest way of raising the necessary money. The State laws as to preferred and common stockholders vary and the by-laws of the company also have something to do with the conditions governing the preferred and common stock. In many cases the preferred stockholders have no vote, while in other cases they have the same voting privileges as the common stockholders. This should be provided for in the articles of incorporation and, if thought necessary, the preferred stockholders may even be given an advantage in voting power. The preferred stock may consist in this:

That it draws the same dividends as the common stock, but it may be provided that each share of preferred stock shall have one full vote and each share of common stock a fractional vote. This will, of course, make the preferred stock more valuable, not only on the start but for the whole existence of the corporation.

It may be that the offer of preferred stock at par will not be sufficient to induce people to invest, and if the showing is not quite good enough to insure this investment it is well to have sufficient treasury stock so that for every share of preferred stock subscribed a share of common stock can be given as a bonus. This gives the subscriber a particularly good showing and it may be necessary and advisable to do this. In this, as in other cases, the patentee or promoter must be governed by circumstances and must gauge his capital and amount of stock to be sold and given as a bonus so as to still leave him, if possible, with the controlling interest, or with an equivalent therefor. In organizing a corporation the capital should be made not too large but still large enough so that the first stock can be sold at less than par. It is a peculiarity of human nature that a person must think that he is getting a bargain and must believe he is buying stock less than par on the start to induce him to invest. We do not know how to account for this. People who are constantly dealing in stocks and know that the ultimate value must depend on what they pay are still affected by the par value—that is to say, if a thing is worth one hundred thousand dollars and is stocked for ten thousand dollars, a man familiar with stocks would know that the stock is worth one thousand dollars a share, but he still would wish to buy for less than par, even if this were one dollar, and if the company is overstocked he does not make allowance for the overstocking if he buys the stock for a good deal less than par. This is a singular fact but it is one borne out by the experience of every person who has had anything to do with organizing and promoting corporations. The organizer should therefore make his capital large enough so that he can cater to this feeling and sell his stock at a substantial reduction from its par value and still realize what he thinks is right.

Bonds.—Instead of raising money by the sale of treasury stock or by providing a treasury fund, it may be preferable to mortgage the property of the company, if it have tangible property, to secure a definite amount of bonds which can then be sold—their price, of course, depending on whether the mortgaged property and the income therefrom is ample to meet the interest and principal represented by the bonds. This bonding is a simple matter and the principles governing it are practically the principles relative to the ordinary bond and mortgage of real or other property, except that the bond instead of issuing to a single person is usually issued by a trustee, to whom the mortgage is given, and instead of being a single bond is made in the form of many bonds. There is no rule governing the par value of bonds, but their denomination is usually one thousand dollars and to make them specially acceptable both principal and interest should be made payable in gold.

Subscription List.—A few parties having agreed to take part in the organization of the corporation and to take a certain amount of stock at a certain price, no subscription list is usually required, but if it is not known how many parties will participate and it is desired to get enough to raise a certain amount of money, and if such an amount is really essential, then it is well to provide a subscription list, which the prospective stockholders may sign and by which they agree to take a certain number of shares of stock at a certain price per share. Under ordinary conditions the price must be par, but as pointed out the stock may be sold at a different price by first paying it in by way of a property transfer and then selling it. If this is to be done the manner in which it is to be done should be set forth in the subscription list. The subscription list should also state under what State or Territory the corporation is to be organized, the total amount of its capital stock, the amount to be paid in and how it is paid in, and all the conditions governing the organization. Everything governing the organizing and starting of the company should be fully and frankly set forth, so that each subscriber may know all the conditions and cannot say that he signed under a misapprehension and cannot accuse any of the parties to the organization of fraudulent or sharp practices. If the organization is to be successful the parties thereto must start with a frank and full understanding of all the circumstances of the case. While usually there are strong temptations to depart from this rule, yet it will be found in the end that it is the only sure and safe one. It is better to start right, even though it be a little harder to make the start. It will be seen, of course, that the form of a subscription list will vary according to circumstances and in the Appendix forms have been given which can be varied to suit most cases.

CHAPTER IV.

THE PROMOTER.

The promoter is one of the new things of the past generation which does not come under the head of patentable subject-matter. A few years since and he was a curiosity; a few years later he was looked upon as a person to be avoided as one would avoid the plague, and finally, in the face of much tribulation, he has come to have a defined and recognized place in the business community. It is true that there are still many promoters, so called, who are of no earthly use, who are unmitigated bores and who are nuisances generally. But the genuine article, the real promoter, is a person of a good deal of ability and of a great deal of use in the world.

This is now generally understood. The promoter who is worthy of the name will not undertake to push to completion a business scheme unless he can see that the scheme is a good one and promises profit to himself, his client and investors. In connection with the term many people will immediately call to mind the famous "Humbug" Hooley, of London, and the havoc he wrought among investors and among the nobility, but, notwithstanding the fact that Hooley and many others have been examples of all that is bad in business, still there are promoters who are honest, capable and valuable members of the business world. There is an old saying to the effect that it is what a man saves that makes him wealthy, and not what he makes, but this doctrine is all right for infants in finance, though on second thought, even the primitive financiers are unable to see how to save out of an ordinary salary or income the amount represented by the colossal fortunes which have been accumulated in this country within the last few years.

A man has time, no doubt, to lay up treasures in heaven during his short career, but if he wishes to get rich he cannot do it by the old-fashioned process, though, no doubt, this is commendable. At the present time wealth is usually made quickly, if at all. A party finds himself in possession of a good thing, and he pushes it, and makes the most of it while he can do so, and before competition cuts his profit. It is necessary to do business on a big scale and rush it to the last limit while it can be done at a profit, and this can be done by means of a large corporation better, perhaps, than in any other way. The connection between this subject and the text, to wit, the promoter, is that many of the eminent financiers and business men of America have first been promoters, and, in fact, many of them are nothing else now, and the successful promoter must be paid, because he succeeds in helping his clients, who therefore are made wealthy by the same process which makes him wealthy. It is doubtless understood that if a stock company be successfully promoted on the lines already laid out it is one of the quickest ways known to make a great deal of money, and it may be perfectly legitimate so far as the inventor, promoter and investors are concerned, for it should be the case that the promoter, by interesting the investors in the company and invention, works a benefit to all parties concerned.

A promoter is one who encourages and carries forward a business scheme with a view of getting capital interested in it, and more especially one who carries forward such a scheme by means of a joint-stock company or corporation. This may be by organizing an original corporation to acquire certain property and afterward manipulate and work it in a business way, or it may be in the organizing of a large company, beginning to be technically known as a trust, for the purpose of absorbing a series of smaller corporations. This is an age of specialties and specialists, and it is found, as a rule, that a skilled specialist can do much better in his line than can a person not so specially skilled, even though the latter may be a very able man. For this reason, in disposing of a patent or a patent interest by means of a joint-stock company or otherwise, the patentee may not be competent or sufficiently skilled to promote the company himself, and it may be to his interest to call in the service of a skilled promoter. In doing this, let him ascertain, first, that the promoter is skilled; secondly, that he is honest. This being done, he will find that the promoter asks so much for his services that he will hesitate about employing them. But, remembering the old adage that "Half a loaf is better than no bread," he will probably make terms with the promoter. In doing this he should be put to no actual expense, because the promoter will not take up the matter proposed unless it is promising, and if this is the case he will ask for a large contingent interest. If the results are to be in cash, he will expect a substantial percentage of the cash, and if a part of the receipts are stock, he will expect a good interest in the stock, and he is worth it. There was a time when investors, if they invested at all, insisted on being the first in an enterprise, unless it was already paying, because they wished to be sure of "getting in on the ground floor," and while this is still true, yet these people recognize the value of the promoter's services, and are willing to consider that the person who formulates the new scheme and carries it forward to success is worthy of his hire, and should receive substantial recognition.

It is generally known that the bulk of the work, either in organizing a new company and interesting capital, or in effecting a combination of established concerns, falls on the promoter, and, as above stated, investors expect to pay for this work. It is this very work which makes the proposed scheme profitable. It is no small undertaking to combine large interests which, for instance, have up to the moment of combination been battling with each other, or to induce large investments in an untried scheme. As an instance of the promoter's recognized value, it is said that a Pittsburg iron manufacturer who in 1898 combined some of the immense iron and steel industries of the country under a single management, made over $2,000,000 out of the operation, and while this, of course, came out of the combination, yet he was well worth the price, because the saving effected was something enormous, and only a skilled man, knowing thoroughly the people with whom he was dealing, could carry the combination into effect.

If a party or parties wish to sell the patent rights by means of a corporation organized for the purpose, and cannot see their way clear to do it themselves, they had better make connections with a good promoter. Promoting a new enterprise is a difficult thing to do. One must be never-tiring in presenting the matter to the right people, and the matter must also be presented rightly. If the scheme is a large one, it is as easy to float as a small one, and perhaps easier, if only the right people can be reached. There is money enough for any enterprise, no matter how gigantic, if it only promises sufficient returns, but the average man will find it easier to get an interview with the President of the United States than with most any well-known capitalist and investor. He will find the man he tries to see so hedged about with offices, office boys, clerks, secretaries ; so many questions asked him concerning his business, and so many obstacles placed in his way that he will, nine times out of ten, give the matter up in despair. But the skilled promoter does not despair. He knows that he has a good thing, and he is going to see it through. He has found out, too, that brass is almost as good as gold if rightly used, and he knows exactly how to use it. He is a gentleman, but he has found out that cheek must be cultivated and used. He knows nearly everybody, and, if necessary, will know the rest of mankind. If it is necessary to interview any man on earth he will manage to interview him. He gets himself invited to dinners, the Lord knows how, and he gets into exclusive sets in some way or other. He comes to a capitalist with an introduction which cannot be ignored, he wriggles and twists and works, and finally gets what he is after.

The capitalist of to-day is the promoter of yesterday. If he has succeeded a few times he ceases to be a promoter except on his own account, but there are many smart, energetic promoters who are in this transitory stage from unknown private citizens to well-known capitalists, and their services can be enlisted in promising enterprises. It is our experience that it often pays to use them, though care must be exercised in their selection, as in everything else. A third person who is really competent, and who has a good-sized interest at stake, can frequently talk up an enterprise better than the prime mover, and when questions begin to reach some definite conclusion, and propositions are being made, and perhaps accepted, it is sometimes an advantage to be able to say that one must see another party before deciding.

Finally, we consider the promoter who is really successful, one of the smartest and most energetic men on earth. He does not lend himself to questionable schemes, but if there is merit in one with which he is connected he is

going to see the matter pushed to a successful
issue, if it is possible to be done. He takes
large risks ; he spends money freely, he loses
nonchalantly, and if he wins he wins large
stakes, and is entitled to them. We commend
him to people who are engaged in carrying forward any new enterprise, unless the prime
mover happens, as is sometimes the case, to be
skilled in this line himself. Let the inventor
make an alliance with a good promoter and
carry through one good scheme to a successful
conclusion, and he has had a liberal education
in financial methods, and will see how money
is made quickly instead of by the old-time processes referred to above ; he may even graduate
into a promoter-inventor, and finally into a
capitalist himself. We are not moralizing on
whether the acquisition of quick wealth is
strictly in accordance with good ethics, but are
simply trying to show the not-over-rich inventor how he can compete with others and
gather in his share of money, and whether or
not the game is worth the candle we leave to
him.

OFFICIAL FEES.

Fees payable to the Patent Office must be paid in advance and upon making application for any action in which a fee is payable.

The following is a schedule of fees for patents, trade-marks, labels, prints, etc.:

On filing each original application for a patent, except in design cases, . $15.00

On issuing each original patent, except in design cases, ... 20.00

In design cases:

For three years and six months, . 10.00

For seven years, ... 15.00

For fourteen years, ... 30.00

On filing each caveat, ... 10.00

On every application for the reissue of a patent, ... 30.00

On filing each disclaimer, . . . 10.00

On an appeal for the first time from the Primary Examiners to the Examiners-in-Chief, ... 10.00

On every appeal from the Examiners-in-Chief to the Commissioner, . . 20.00

For certified copies of printed patents:

For specification and drawing, per copy, ...05

For the certificate, ...25

For the grant, ...50

For certifying to the duplicate of a model, ...50

For manuscript copies of records, for
every one hundred words or frac-
tion thereof, ... $0.10

If certified, for the certificate, additional, ...25

For uncertified printed copies of the
specifications and accompanying
drawings of patents, each, . . .05

When ordered by subclasses, each, . .03

When ordered by classes, each, . . .02

And when the entire set of all patents
granted is ordered, each, ...01

For the drawings, if in print, ...05

For copies of drawings not in print the
reasonable cost of making them.

For recording every assignment, agreement, power of attorney or other
paper, of three hundred words or
under, ... 1.00

Of over three hundred and under one
thousand words, ... 2.00

Of over one thousand words, . . 3.00

On filing an application for registration
of a trade-mark, ... 25.00

On filing an application for registration
of a label, ... 6.00

On filing an application for registration
of a print, ... 6.00

APPENDIX

APPENDIX

In the Appendix is given a few of the forms which are more often used, and while any specific form is not essential still it is well to follow the custom in this respect as an instrument of the usual character can be more readily construed.

No specific form of acknowledgment is given on the several forms of assignment because this varies with different States and it is well to follow the form used in the State where the instrument is executed.

It will be noticed that in the assignment before issue, a request is made to the Commissioner of Patents to have the patent issue to the assignee. This is not essential, but is usually advisable. If, however, there is any reason why it should not so issue, the clause making the request can be omitted.

ASSIGNMENT OF AN ENTIRE INTEREST IN AN INVENTION BEFORE THE ISSUE OF LETTERS PATENT.

Whereas, I, John Jones, of the City, County and State of New York, have invented certain new and useful Improvements in Can Openers for which I am about to make application for Letters Patent of the United States, which application I have signed and executed this......day of......189.. ; and

Whereas, James Smith, of Boston, in the County of Suffolk and State of Massachusetts, is desirous of acquiring the entire interest in the aforesaid invention and in the Letters Patent to be issued therefor ;

Now, Therefore, To all whom it may concern, be it known that for and in consideration of the sum of ......dollars and other valuable considerations to me in hand paid, the receipt of which is hereby acknowledged, I, the said John Jones, have sold, assigned and transferred, and by these presents do sell, assign and transfer unto the said James Smith the whole right, title and interest in and to the aforesaid invention, as set forth in the application above referred to, and in and to the Letters Patent which may issue therefor, and I hereby authorize and request the Commissioner of Patents to issue the said Letters Patent to the said James Smith as the Assignee of my entire right, title and interest therein, for the sole use and behoof of the said James Smith and his legal representatives.

In Testimony Whereof, I have hereunto set my hand and affixed my seal this......day of......189...
....

Witnesses :
....
....   (Acknowledgment.)

If the assignee is a corporation the preamble would be as follows :

" *Whereas*, I, John Jones, of the City, County and State of New York, have invented certain new and useful Improvements in Can Openers for which I am about to make application for Letters Patent of the United States, which application I have signed and executed this......day of......189.. ; and

" *Whereas*, the Nineteenth Century Manufacturing Company, a corporation organized and existing under the laws of the Territory of Arizona and doing business in the City, County and State of New York, is desirous of acquiring the entire interest in the same " ;

The remainder will be substantially like the body of the assignment already given.

If both parties to the transfer are corporations, the preamble will, of course, set forth where each company does business and under what State or Territory it is organized.

ASSIGNMENT OF THE ENTIRE INTEREST IN
LETTERS PATENT.

If the patent has already issued it should be identified by title, number and date, and the assignment may then be as follows:

Whereas, I, John Jones, of the City, County and State of New York, did obtain Letters Patent of the United States for an Improvement in Can Openers, which Letters Patent are numbered......and bear date the twenty-fourth day of January, 1899 ; and

Whereas, I am now the sole owner of the said patent and of all rights under the same ; and

Whereas, James Smith, of Boston, in the County of

Suffolk and State of Massachusetts, is desirous of acquiring the entire interest in the same ;

Now, Therefore, To all whom it may concern, be it known that for and in consideration of the sum of ......dollars and other valuable considerations to me in hand paid, the receipt of which is hereby acknowledged, I, the said John Jones, have sold, assigned and transferred, and by these presents do sell, assign and transfer unto the said James Smith the whole right, title and interest in and to the said Improvement in Can Openers and in and to the Letters Patent therefor aforesaid, the same to be held and enjoyed by the said James Smith for his own use and behoof and for the use and behoof of his legal representatives to the full end of the term for which said Letters Patent are granted as fully and entirely as the same would have been held and enjoyed by me had this assignment and sale not been made.

In Testimony Whereof, I have hereunto set my hand and affixed my seal this......day of......189...

JOHN JONES. (Seal.)

Witnesses:
    A. B.
    C. D.   (Acknowledgment.)

ASSIGNMENT OF APPLICATION AFTER FILING.

If the patent has not been issued but the application has been filed, the form would be substantially like the above, except that instead of quoting the number and date of the patent, the serial number and date of filing should be given, and there can be a request to the Commissioner of Patents, as in form 1, to issue the patent to the assignee.

ASSIGNMENT OF AN UNDIVIDED INTEREST IN LETTERS PATENT.

If the assignee is to have an undivided interest in the invention and patent, the form will be as follows:

Whereas, I, John Jones, of the City, County and State of New York, did obtain Letters Patent for an Improvement in Can Openers, which Letters Patent are numbered......and bear date the twenty-fourth day of January, 1899; and

Whereas, James Smith, of Boston, County of Suffolk and State of Massachusetts, is desirous of acquiring an interest in the same;

Now, Therefore, To all whom it may concern, be it known that for and in consideration of the sum of ......dollars and other valuable considerations to me in hand paid, the receipt of which is hereby acknowledged, I, the said John Jones, have sold, assigned and transferred, and by these presents do sell, assign and transfer unto the said James Smith an undivided one-half part of my whole right, title and interest in and to the said invention and in and to the Letters Patent therefor aforesaid, the said undivided one-half part to be held and enjoyed by the said James Smith for his own use and behoof and for the use and behoof of his legal representatives to the full end of the term for which said Letters Patent are granted as fully and entirely as the same would have been held and enjoyed by me had this assignment and sale not been made.

In Testimony Whereof, I have hereunto set my hand and affixed my seal this......day of......189...

JOHN JONES. (Seal.)

Witnesses:
  A. B.
  C. D.   (Acknowledgment.)

GRANT OF A TERRITORIAL INTEREST AFTER THE ISSUE OF PATENTS.

Whereas, I, John Jones, of the City, County and State of New York, did obtain Letters Patent of the United States for an Improvement in Can Openers, which Letters Patent are numbered..... and bear date the twenty-fourth day of January, 1899 ; and

Whereas, I am now the sole owner of the said patent and all rights under the same in the territory hereinafter mentioned; and

Whereas, James Smith, of Boston, County of Suffolk and State of Massachusetts, is desirous of acquiring an interest in the same ;

Now, Therefore, To all whom it may concern, be it known that for and in consideration of the sum of ......dollars and other valuable considerations, to me in hand paid, the receipt of which is hereby acknowledged, I, the said John Jones, have sold, assigned and transferred, and by these presents do sell, assign and transfer unto the said James Smith all the right, title and interest in and to the said invention as secured to me by the Letters Patent above referred to, for, to and in the States of Maine and New Hampshire, and the Counties of Essex and Suffolk in the State of Massachusetts, and for, to or in no other place or places, the same to be held and enjoyed by the said James Smith within and throughout the territory mentioned, but not elsewhere, for his own use and behoof and for the use and behoof of his legal representatives, &c.

....

Witnesses :

....

....   (Acknowledgment.)

SIMPLE FORM OF SHOP-RIGHT LICENSE.

In consideration of......dollars, to me in hand paid by the Ajax Manufacturing Company, a corporation organized and existing under the laws of the State of Pennsylvania and doing business in Philadelphia in said State, I do hereby license and empower the said Ajax Manufacturing Company to manufacture in said City of Philadelphia the Improvement in Sole Edge Burnishing Machines, for which Letters Patent of the United States numbered......were granted to me on the seventeenth day of January, 1899, and to sell the machine so manufactured throughout the territory of the United States to the full end of the term for which said Letters Patent are granted.

Signed at the City and County of Philadelphia, State of Pennsylvania, this first day of February, 1899.

A. B.

Witnesses :
  C. D.
  E. F.

FORM OF LICENSE WITH ROYALTY.

This Agreement, made this eighth (8th) day of February, in the year one thousand eight hundred and ninety-nine (1899) by and between A. B., of Washington, District of Columbia, party of the first part, and C. D., of Pittsburgh, Allegheny County, Pennsylvania, party of the second part, *Witnesseth :*

The party of the first part did on the seventh (7th) day of February, 1899, obtain Letters Patent of the United States for an Improvement in Rock Drills, which Letters Patent are numbered......, and the party of the second part is desirous of entering into the manufacture of said Rock Drills ;

Now, Therefore, In view of the premises, and in consideration of the sum of one dollar by each to the other paid, the parties have agreed as follows :

First: The party of the first part hereby grants to the party of the second part the full and exclusive right to make, use and vend the aforesaid invention, referred to in the Letters Patent above named, throughout all that part of the territory of the United States lying east of the Mississippi River, including the State of Minnesota, and in no other places, to the end of the term for which said Letters Patent were granted, subject, however, to the conditions hereinafter named.

Second: The party of the first part further agrees to protect the Letters Patent herein referred to, to proceed at his expense against infringers of the said patent and to defend the party of the second part if he should be sued for infringement, if said alleged infringement consists in the manufacture of the Rock Drill herein referred to.

Third: The party of the second part, as a guaranty that this contract will be carried out in good faith, has paid to the party of the first part the sum of one thousand dollars, the receipt of which is hereby expressly acknowledged by the party of the first part, and the party of the second part further agrees to pay to the party of the first part the sum of ten dollars for each rock drill of the kind specified above manufactured and sold by him, such payments to be made every sixty days and within ten days from the time of making returns as specified below.

Fourth: The party of the second part agrees to manufacture and sell not less than twelve hundred machines in any one year and agrees to make full and true returns under oath on the first days of January, March, May, July, September and November of each year of the number of Rock Drills made and sold by him and agrees further that his books of sales shall be open at all reasonable times to the inspection of the party of the first part, and within ten days after each return day the party of the second part agrees to pay to the party of the first part the amount of royalties due.

Fifth : Upon the failure of the party of the second part to make returns or to make payment of license fees, as herein provided, for more than thirty days after the dates herein named, the party of the first part may terminate this license by serving on the party of the second part a written notice to that effect, but the party of the second part shall not thereby be discharged from any liability to the party of the first part for any license fees due at the time of the service of said notice, and of the one thousand dollars paid by the party of the second part no part of such sum shall be considered as being applied on royalties herein mentioned.

In Witness Whereof, the parties above named have hereunto set their hands and seals the day and year first above written at Pittsburg, in the County of Allegheny and State of Pennsylvania.

A. B.

Witnesses :   C. D.

  E. F.

  G. H.

SUBSCRIPTION AGREEMENT BEFORE ORGANIZA-
TION OF A CORPORATION.

Whereas, John Jones, of the City, County and State of New York, is the owner of Letters Patent of the United States No......., dated February 7th, 1899, for Improvement in Sewing Machines ; also of Letters Patent of the United States No......., dated January 24th, 1899, for an Improvement in a Pressure Foot for Sewing Machines, and

Whereas, the said John Jones proposes to organize a corporation under the General Laws of the State of ......, to be known as the Jones Manufacturing Company, with a capital stock of one hundred thousand ($100,000.00) dollars, divided into shares of a par value of one hundred ($100.00) dollars each. Said corporation to be for the purpose of engaging in the manufacturing of machinery and tools of all kinds and especially of the maufacture of sewing machines and pressure feet under the above-named patents, but also to do any business within the general scope of a manufacturing company, and it is desired by the undersigned to become a shareholder in the above corporation ;

Now, Therefore (insert name of subscriber), does hereby promise and agree to, and with the said John Jones, in consideration of the promises of the said John Jones hereinafter stated, that he will pay to the said John Jones or to any person or corporation to whom he may assign this agreement, on demand, the sum of......dollars, being the subscription price of ......shares of the capital stock of the said corporation, or such part of said subscription price as may be called for. The stock thus paid for to be delivered at the earliest possible moment after the organization of the company, and meanwhile proper receipts or scrip to be issued to the undersigned.

This Agreement is conditioned as follows :

First : The said John Jones shall procure other bona fide subscriptions aggregating in all not less than twenty-five thousand ($25,000.00) dollars of the capital stock of the said corporation on the same terms as stated herein.

Second : On the organization of said corporation, the said John Jones shall make an assignment to the said Jones Manufacturing Company by which he shall set over to the said company the whole right, title and interest in and to the Letters Patent for sewing machines and pressure feet herein named.

Third : The said John Jones shall accept in payment for the said patents fifty thousand ($50,000.00) dollars of the capital stock of the Jones Manufacturing Company herein named.

Fourth : The said John Jones, on his part, in consideration of the foregoing, promises to use his best endeavors to obtain the said twenty-five thousand ($25,000.00) dollars of subscriptions and his best efforts to perfect the organization of the said corporation.

Witness our hands and seals this......day of......
1899, at the City, County and State of New York.

..........................(Seal.)

..........................(")

..........................(")

If the corporation is to have a certain amount of preferred stock the fact should be set forth in the agreement and the character of such preferred stock described, that is to say, if it is to be six per cent. cumulative stock or six per cent. non-cumulative, or whatever the nature of its preferment—the matter should be distinctly set out in the agreement. In fact, any pertinent matter to the organization should be clearly set forth in the agreement so that there can be no misunderstanding between the parties. Obviously this agreement will differ greatly with different cases, but its general tenor will enable a skilled person to use it to meet the exigencies of almost any case.

The form of a certificate will vary somewhat in different States, but we present two forms; one for any stock and one providing for preferred stock, which are adapted to meet the requirements of the New Jersey Corporation Law, and these can easily be changed to conform to the laws of other States or Territories.

Form having no preferred stock:

CERTIFICATE OF INCORPORATION

OF THE

(Here insert name of company.)

First : The name of the Corporation is (here insert company's name.)

Second : The location of its principal office in the State of New Jersey is at No.......Street, in the city of Jersey City, County of Hudson. The name of the agent therein and in charge thereof upon whom process against this Corporation may be served is (here insert name of agent.)

Third : The objects for which, and for any of which, the Corporation is formed are to do any or all of the things herein set forth, to the same extent as natural persons might or could do and in any part of the world, to wit: (here enumerate the special objects of the corporation.)

In Furtherance of, and not in limitation of, the general powers conferred by the Laws of the State of New Jersey, it is hereby expressly provided that the Company shall have also the following powers :

To manufacture, purchase or otherwise acquire, to hold, own, mortgage, pledge, sell, assign and transfer, or otherwise dispose of, to invest, trade, deal in and deal with goods, wares and merchandise and property of every class and description.

To acquire the good will, rights and property, and to undertake the whole or any part of the assets and liabilities, of any person, firm, association or corporation, and to pay for the same in cash, stock of this Company, bonds or otherwise.

To apply for, purchase, or otherwise acquire, and to hold, own, use, operate, and to sell, assign, or to otherwise dispose of, to grant licenses in respect of or otherwise turn to account any and all inventions, improvements and processes used in connection with, or secured under Letters Patent of the United States or elsewhere, or otherwise, and with a view to the working and development of the same to carry on any business, whether manufacturing or otherwise, which the Corporation may think calculated directly or indirectly to effectuate these objects.

To enter into, make, perform and carry out contracts of every kind with any person, firm, association, corporation, without limit as to amount, to draw, make, accept, endorse, discount, execute and issue promissory notes, bills of exchange, warrants, bonds, debentures, and other negotiable or transferable instruments.

To have one or more offices, to carry on all or any of its operations in business, and without restriction to the same extent as natural persons might or could do, to purchase or otherwise acquire, to hold, own, to mortgage, sell, convey, or otherwise dispose of, without limit as to amount, real and personal property of every class and description, in any state or territory of the United States, and in any foreign country or place.

In General to carry on any other business in connection therewith, whether manufacturing or otherwise, and with all the powers conferred by the Laws of New Jersey on corporations under the act hereinafter referred to.

The duration of the Corporation shall be unlimited.

Fourth : The total authorized capital stock of this

corporation is......dollars ($......), divided into
...... (......) shares of one hundred dollars ($100)
each.

Fifth : The names of the incorporators (the post
office address of each is No. ....Street, Jersey
City, New Jersey), and the number of shares subscribed for by each, the aggregate of which ($......)
is the amount of capital with which the Company will
commence business, are as follows :

Name......P. O. address......No. of shares......

Sixth : The Board of ·Directors shall have power,
without the assent or vote of the stockholders, to
make, alter, amend and rescind the by-laws of this
corporation, to fix the amount to be reserved as a
working capital, to authorize and cause to be executed
mortgages and liens without limit as to amount upon
the real and personal property of this Corporation.

With the consent in writing and pursuant to the
vote of the holders of a majority of the stock issued
and outstanding, the Directors shall have power and
authority to sell, assign, transfer or otherwise dispose
of the whole property of this Corporation.

The Directors shall from time to time determine
whether and to what extent, and at what times and
places and under what conditions and regulations the
accounts and books of the Corporation, or any of them,
shall be open to the inspection of the stockholders ;
and no stockholder shall have any right of inspecting
any account or book or document of the Corporation
except as conferred by statute or authorized by the
Directors, or by resolution of the stockholders.

The Board of Directors, in addition to the powers
and authorities by statute and by the by-laws expressly conferred upon them, may exercise all such
powers and do all such acts and things as may be
exercised or done by the corporation, but subject,
nevertheless, to the provisions of the statute, of the
charter, and to any regulations that may from time to

time be made by the stockholders, provided that no regulations so made shall invalidate any provisions of this charter, or any prior acts of the Directors which would have been valid if such regulations had not been made.

The Corporation may in its By-laws confer powers additional to the foregoing upon the Directors, and may prescribe the number necessary to constitute a quorum of its Board of Directors, which number may be less than a majority of the whole number.

The Board of Directors may, by resolution passed by a majority of the whole Board, designate two or more of their number to constitute an Executive Committee, which Committee shall for the time being, as provided in said resolution or in the By-laws of said Corporation, have and exercise all the powers of the Board of Directors in the management of the business and affairs of the Company, and have power to authorize the seal of the Corporation to be affixed to all papers which may require it.

The Directors shall have power to hold their meetings, to have one or more offices, and to keep the books of the Corporation (except the stock and transfer books) outside of this State, at such places as may be from time to time designated by them.

It is the intention that the objects specified in the third paragraph shall, except where otherwise expressed in said paragraph, be nowise limited or restricted by reference to or inference from the terms of any other clause or other paragraph in this Charter, but that the objects specified in each of the clauses of this paragraph shall be regarded as independent objects.

The Undersigned, for the purpose of forming a Corporation in pursuance of an Act of the Legislature of the State of New Jersey, entitled "An Act Concerning Corporations" (Revision of 1896), and the various acts amendatory thereof and supplemental thereto, do make, record and file this certificate, and do respectively agree to take the number of shares of stock hereinbefore set forth, and accordingly hereunto have set our hands and seals.

Dated, Jersey City, N. J.,

    In presence of

.......................(Seal.)

.......................(")

.......................(")

.......................(")

State of........ } ss. :

County of...... }

Be it remembered that on this......day of......A. D. eighteen hundred and ninety-......before the undersigned personally appeared......who I am satisfied are......the persons named in and who executed the foregoing certificate, and I having first made known to them and each of them the contents thereof, they did each acknowledge that they signed, sealed and delivered the same as their voluntary act and deed.

....

Received in the Hudson County, N. J., Clerk's Office,......189.., and recorded in the Clerk's Record No.......on Page......

.................Clerk.

Endorsed "Filed......189..

    GEORGE WURTS, Secretary of State."

Under Article Third, which specifies the objects of the corporation, care should be taken to give the company sufficient scope to do any business it may ever wish to do, even though the business is not contemplated at the time of organization.

The form above given can be varied to suit different States as stated above, but in very many States blanks can be obtained from the Secretary of State or from dealers in law blanks, which are suitable for the purpose.

PREFERRED AND COMMON STOCK.

CERTIFICATE OF INCORPORATION

10-cent internal
revenue stamp
cancelled.

OF THE

(Here insert name of company).

First : The name of the Corporation is the (here insert company's name).

Second : The location of the principal office in the 'State of New Jersey is at No. .. .. Street, in the City of Jersey City, County of Hudson. The name of the agent therein and in charge thereof, upon whom process against this corporation may be served, is (herein insert name of agent).

Third : The objects for which and for any of which, the corporation is formed, are to do any or all of the things herein set forth, to the same extent as natural persons might or could do and in any part of the world, to wit : (here enumerate the special objects of the corporation.)

In furtherance of, and not in limitation of, the general powers conferred by the Laws of the State of New Jersey, it is hereby expressly provided that the company shall have also the following powers :

To do any or all of the things herein set forth, to the same extent as natural persons might or could do, and in any part of the world.

To manufacture, purchase or otherwise acquire, to hold, own, mortgage, pledge, sell, assign and transfer, or otherwise dispose of, to invent, trade, deal in and deal with goods, wares and merchandise and property of every class and description.

To acquire the good will, rights and property, and to undertake the whole or any part of the assets and liabilities, of any person, firm, association or corporation, and to pay for the same in cash, stock of this company, bonds or otherwise.

To apply for, purchase, or otherwise acquire, and to hold, own, use, operate, and to sell, assign, or to otherwise dispose of, to grant licenses in respect of or otherwise turn to account any and all inventions, improvements, and processes used in connection with, or secured under, Letters Patent of the United States or elsewhere, or otherwise, and with a view to the working and development of the same to carry on any business, whether manufacturing or otherwise, which the corporation may think calculated directly or indirectly to effectuate these objects.

To enter into, make, perform and carry out contracts of every kind with any person, firm, association or corporation, and, without limit as to amount, to draw, make, accept, endorse, discount, execute and issue promissory notes, bills of exchange, warrants, bonds, debentures and other negotiable or transferable instruments.

To have one or more offices, to carry on all or any of its operations and business, and without restriction to the same extent as natural persons might or could do, and to purchase or otherwise acquire, to hold, own, to mortgage, sell, convey or otherwise dispose of, without limit as to amount, real and personal property of every class and description, in any State, Territory or Colony of the United States, and in any foreign country or place.

To do any or all the things herein set forth to the same extent as natural persons might or could do, and in any part of the world, as principals, agents, contractors, trustees or otherwise.

In general, to carry on any other business in connection therewith, whether manufacturing or otherwise, and with all the powers conferred by the Laws of New Jersey upon corporations under the Act hereinafter referred to.

The duration of the corporation shall be unlimited.

Fourth : The total authorized capital stock of this corporation is......dollars, ($......) divided into ......(......) shares of one hundred dollars ($100) each.

Of said stock......shares shall be preferred stock, and the balance, ...shares, shall be common or general stock.

Said preferred stock shall entitle the holder thereof to receive out of the net earnings, and the company shall be bound to pay a fixed yearly cumulative dividend at the rate of but not exceeding......per centum, payable......before any dividend shall be set apart or paid on the common stock. Such preferred stock shall, at the discretion of the company, be subject to redemption at par on......19...., or on any dividend day thereafter.

Such preferred stock may be issued as and when the Board of Directors shall determine.

The holders of preferred stock shall, in case of liquidation or dissolution of the company, be entitled to be paid in full, before any amount shall be paid to the holders of the general or common stock.

The holders of such preferred stock may choose ...... of the Directors, and the remainder of the Board shall be chosen by the common or general stockholders.

Fifth : The names of the incorporators (the post-office address of each is No. .. .. Street, Jersey

City, N. J.), and the number of shares subscribed for by each, the aggregate of which ($......) is the amount of capital with which the Company will commence business, are as follows :

Name   P. O. Address   Number of shares

....

Sixth : The Board of Directors shall have power, without the assent or vote of the stockholders, to make, alter, amend and rescind the By-Laws of this Corporation, to fix the amount to be reserved as working capital, to authorize and cause to be executed mortgages and liens without limit as to amount upon the real and personal property of this Corporation.

With the consent in writing and pursuant to the vote of the holders of a majority of the stock issued and outstanding, the Directors shall have power and authority to sell, assign, transfer or otherwise dispose of the whole property of this Corporation.

The Directors shall from time to time determine whether and to what extent, and at what times and places and under what conditions and regulations, the accounts and books of the Corporation, or any of them, shall be open to the inspection of the stockholders; and no stockholder shall have any right of inspecting any account or book or document of the Corporation, except as conferred by statute or authorized by the Directors, or by a resolution of the stockholders.

The Board of Directors, in addition to the powers and authorities by statute and by the By-Laws expressly conferred upon them, may exercise all such powers and do all such acts and things as may be exercised or done by the Corporation, but subject, nevertheless, to the provisions of the statute, of the Charter, and to any regulations that may from time to time be made by the stockholders, provided that no regulations so made shall invalidate any provisions of this Charter, or any prior acts of the Directors which would have been valid if such regulations had not been made.

The Corporation may in its By-Laws confer powers additional to the foregoing upon the Directors, and may prescribe the number necessary to constitute a quorum of its Board of Directors, which number may be less than a majority of the whole number.

The Board of Directors may, by resolution passed by a majority of the whole Board, designate two or more of their number to constitute an Executive Committee, which Committee shall for the time being, as provided in said resolution or in the By-Laws of said Corporation, have and exercise all the powers of the Board of Directors in the management of the business and affairs of the Company, and have power to authorize the seal of the Corporation to be affixed to all papers which may require it.

Neither the Directors nor the members of the Executive Committee nor the President nor Vice-President shall be subject to removal during their respective terms of office except for cause, nor shall their terms of office be diminished during their tenure.

The Directors shall have power to hold their meetings, to have one or more offices, and to keep the books of the Corporation (except the stock and transfer books) outside of this State, at such places as may be from time to time designated by them.

It is the intention that the objects specified in the third paragraph shall, except where otherwise expressed in said paragraph, be nowise limited or restricted by reference to or inference from the terms of any other clause or other paragraph in this Charter, but that the objects specified in each of the clauses of this paragraph shall be regarded as independent objects.

We, the undersigned, for the purpose of forming a Corporation in pursuance of an Act of the Legislature of the State of New Jersey, entitled "An Act Concerning Corporations" (Revision of 1896), and the various acts amendatory thereof and supplemental thereto, do make, record and file this certificate, and do respectively agree to take the number of shares of stock hereinbefore set forth, and accordingly have hereunto set our hands and seals.

Dated Jersey City, N. J.,
   In the presence of

....

                    ....(Seal.)
                    ....(")
                    ....(")
                    ....(")

State of............ } ss.:
County of.........

Be it remembered that on this ...... day of ...... A. D. ...... before the undersigned personally appeared ...... who I am satisfied are ...... the persons named in and who executed the foregoing certificate, and I having first made known to them, and each of them, the contents thereof, they did each acknowledge that they signed, sealed and delivered the same as their voluntary act and deed.

10-cent internal
revenue stamp
cancelled.

   Received in the Hudson County, N. J., Clerk's Office, ... 1...., ... and recorded in Clerk's Record No. ...... on Page ......

                    .... Clerk.

   Endorsed "Filed
                   .... Secretary of State."

If the stock is to be non-cumulative the word "non" can be inserted before "cumulative."

ASSIGNMENT OF SUBSCRIPTION.

Know all men by these presents,
   That I, ... in consideration of One

Dollar, lawful money of the United States, to me paid before the ensealing and delivery of these presents, the receipt whereof is hereby acknowledged, and for other good and valuable considerations, have sold, assigned, transferred and set over, and by these presents do sell, assign, transfer and set over unto ...... my right, title and interest as a subscriber to and an incorporator of the ...... Company, a corporation organized under the Laws of the State of New Jersey, to the extent of ...... shares, and I do hereby request and direct the said Company to issue the certificate for said ...... shares to and in the name of said .. .... or such other person as he may name.

In witness whereof, I have hereunto set my hand and seal this ...... day of ......, 189..

[L.S.]

Sealed and delivered in the presence of ....

INDEX.

www.ingramcontent.com/pod-product-compliance
Lightning Source LLC
Chambersburg PA
CBHW021948220326
41599CB00012BA/1414